CAP GUNS

WITH VALUES

JAMES L. DUNDAS

77 Lower Valley Road, Atglen, PA 19310

DEDICATION

This book is dedicated to my brother, and all of those brothers and sisters who were killed in Vietnam.

Brother

*Only memories of you my brother have I left to ease the pain,
only memories as of sunshine that is banished by the rain.
With the autumn flowers you faded and with them have gone to rest.
Forgive me in my anguish, for perhaps thou knowest best.
All in vain I try to fathom why you went to Vietnam,
was your gallant soul concealing, what we know so well today?
That you were a true American and believe in the American way,
and now have gone away.
Winter's winds will sigh in mourning, spring will bring the flowers once more,
to embellish summer's grandeur, then to die at autumn's door.
Some day God will call to me, and I will set forth,
but perhaps it's a sunny morning and I will find a garden fair,
full of flowers and little children, and my Brother will be there.
Smiling he will come to meet me, and I think I hear him say.
"Did you say that you had missed me? Why I haven't been away."*

*For Sgt. Jerry R. Dundas
Killed at Khesanh, Vietnam on 5/5/68
from his brother Professor James L. Dundas*

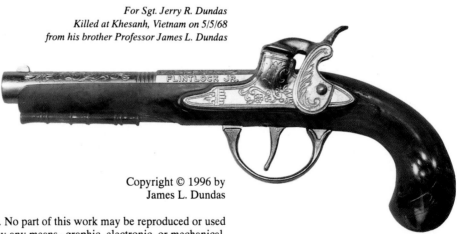

Copyright © 1996 by
James L. Dundas

All rights reserved. No part of this work may be reproduced or used in any forms or by any means--graphic, electronic, or mechanical, including photocopying or information storage and retrieval systems--without written permission from the copyright holder.

Printed in Hong Kong
ISBN: 0-88740-960-1

Published by Schiffer Publishing Ltd.
77 Lower Valley Road
Atglen, PA 19310
Please write for a free catalog.
This book may be purchased from the publisher.
Please include $2.95 for shipping.
Try your bookstore first.

Library of Congress Cataloging-in-Publication Data

Dundas, James L.
    Cap guns : with values / James L. Dundas.
      p. cm.
    ISBN 0-88740-960-1 (paper)
    1. Toy pistols--Collectors and collecting. I. Titl
TS532.4.D84 1996
688.7'28--dc20   95-45515
        CIP

CONTENTS

Preface 5
History 6
Cap Guns
    Cast Iron Single Shot 7
    Cast Iron Automatic 59
    Die-Cast 84
Rifles 140
Holsters 144
Using Patent Numbers 160

Front Cover Photographs

Center: Long Tom, 10 3/8" automatic with a steel cylinder, made of nickel-plated cast iron in 1939 by Kilgore. Shoots disc caps. $250-350
Bottom center: Atomic Disintegrator, 8"-long automatic made of die-cast by Hubley. $300-400
Bottom left: No Name, train 4 7/8"-long single shot made of cast iron by Kenton. $400-600

Back Cover Photographs

Left gun: Flintlock Jr, 7 1/2"-long single shot made of nickel-plated die-cast by Hubley. $30-40
Right gun: Lasso Em Bill, 9"-long single shot made of nickel-plated cast iron with jewels in the grip, made in 1930 by Kenton. $250-300
Holster #1 (top): No Name, 10" long, made of leather by unknown. $40-100
Holster #2: Western Pioneer set, 11" long, made of leather. $80-100
Holster #3: Stallion 45 set, 12 1/2" long, made of leather. $125-150
Holster #4: No Name, 9" long, made of leather by unknown. $80-100

Title Page Photographs

Row 1
Left: No Name, 8" long, made of leather by unknown. $20-30
Center: Derringer, 8 1/2"-long single shot made of plastic and die-cast by Marx. $30-40
Right: Sportsman, 5 3/4"-long automatic made of painted cast iron in 1925 by Kilgore. $40-60

Row 2
Left: Dandy, 2 7/8"-long single shot made of cast iron in 1887. $150-175
Center: Daisy, 4 3/4"-long single shot made of cast iron in 1875 by Stevens. $150-250
Right: Ranger, 9" automatic made of nickel-plated cast iron in 1940 by Kilgore. $100-150

Row 3
Left: C-Boy, 7"-long automatic made of nickel-plated cast iron in 1940 by Kenton. $80-100
Center: Whoopee Bird, 7 1/4"-long single shot made of tin. $100-150
Right: No Name, 4 7/8"-long single shot made of cast iron in 1882 by Ives. $150-200

Row 4
Left: Master, 4 5/8"-long automatic made of nickel-plated cast iron in 1922 by Andas. $60-80
Center: National, model 175 7 3/4"-long automatic made of painted die-cast by National Playthings. $30-40
Right: No Name, 3 3/4"-long single shot made of cast iron in 1878 by Stevens. $150-200

ACKNOWLEDGMENTS

A very special thanks to Tom Kolomjec of Mount Clemens, Michigan, for his knowledge and the use of his tremendous cap gun collection. His enthusiasm for this book was a great help.

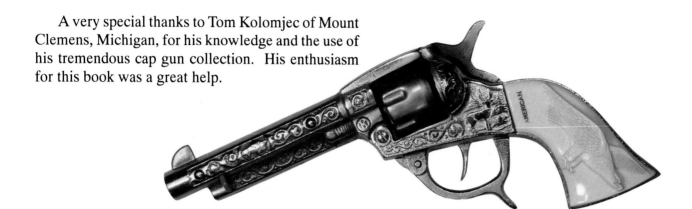

PREFACE

Cap gun collecting is one of the fastest growing collectible fields of the 1990s. Cap guns that sold for a couple of dollars in the 1940s and 1950s are now worth as much as $100 or more, and it looks as if the prices are only going up. Why all the interest in cap guns?

There has always been interest in antique toys, especially cast iron toys like banks, but now they cost hundreds or even thousands of dollars.

While there has been an interest in cast iron cap guns for some time, only now are they coming up in price. Cap guns from the late 1800s through the 1940s are sought for investment purposes as well as for fun. Animated cap guns from this era are especially interesting, with animals, people, or other figures that move to hit the cap (for example, a train that hits a barracuda). These animated guns were made by Ives, Lockwood & Stevens, and Kenton. They are very hard to find, and bring top prices.

Die-cast cap guns from the late 1940s through the 1970s are eagerly sought after for nostalgic reasons—remember the smell of cap gun smoke from when you shot them off as a kid? Baby boomers who were born after 1945 have begun collecting the toys of their childhoods, and cap guns are at the forefront. Some of these die-cast cap guns are becoming very valuable.

James L. Dundas
Professor, Macomb Community College
Warren, Michigan

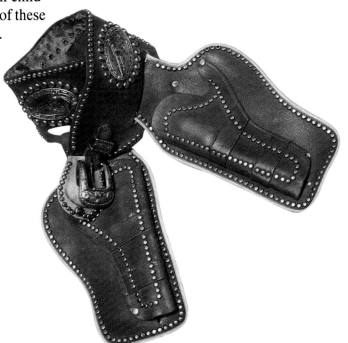

HISTORY

Toy guns were first mass-produced in the 1860s, after the Civil War. Most of these early guns were made of wood and trimmed with metal hardware, although a few were made of lead and iron.

Paper caps had been developed just before the Civil War, and by 1870 they were being used in toy guns. These new cap guns made a loud noise but were fairly harmless. The demand for them was high. Manufacturers worked to keep up with this demand and to produce new designs for this market. Thus, the era from 1870 to 1900 in America is known as the golden age of the toy gun, particularly the cap gun.

By 1880 the cast iron cap gun was the most popular toy gun; manufacturers outdid themselves producing unusual designs. These guns were very fancy but not realistic looking. They were often covered with designs, frequently of the leaf and scroll variety. Some guns had two- and three-dimensional figures; others were animated with moving figures. The animated cap guns are avidly sought after by collectors. 'Head pistols' were made with a human or animal head at the back of the barrel or breach; the cap was placed in the open mouth. These guns are so highly prized by collectors that they seldom come on the market.

By the 1920s these fancy, decorated cap guns had somewhat disappeared. They were replaced by realistic models of real guns. This realism continued through the 1940s. Cast iron was used to make most early toy guns up until World War II, when cast iron was needed for the war effort. During the war, various substitute materials such as wood and composition were used to meet the high demand. After the war, manufacturers turned to steel and die-cast zinc which were less expensive to produce. Plastic also began to be used. Since 1950, most toy cap guns continue to be made of plastic or die-cast material.

Collectors have long searched for cast iron toy guns, but the die-cast cap guns from 1940-1965 are currently the most popular. During this period after World War II, the public was fascinated with cowboys; western movies, radio and T.V. programs, comic books, clothing, and toy guns all evolved from this theme. This era produced many cowboy heroes, including Roy Rogers, Gene Autry, Hopalong Cassidy, and the Lone Ranger. The glamorous guns of this period featured character names, shiny finishes, fancy plastic grips, and holster sets with sparkling stones and studs.

Its easy to see why so many people want to collect guns from this era!

Interest in toy guns has grown tremendously in the past few years as baby boomers search for their childhood memories. Demand far exceeds what is currently available for purchase. Since toy cap guns were breakable, very few have survived. As a result of their heavy use by active children at play, most were damaged, lost, or discarded. Some cap guns found today will have cracked, broken or missing parts. These guns may still be worth something, at least for parts. Guns can still be found in mint condition, sometimes even with the original box.

In this book, the prices listed for each cap gun refer to guns in average working condition. Average condition means a properly working gun with most of its original finish and no missing or broken parts. Cap guns found in mint condition are worth up to 50% more. Original boxes in good condition add 25% to the value of the gun.

Many die-cast cap guns are very expensive or difficult to repair. Be careful when buying a damaged gun. Fancy Western leather holster sets are presently very popular; double holster sets are worth more than single ones. Sets with character names and ornate decorations bring higher prices; the amount of decoration increases the price.

CAST IRON SINGLE SHOT CAP GUNS

Ace, 4 1/8"-long single shot made of cast iron in 1925 by Kilgore. $30-40

Acorn, Pat. June 17 1890 3 1/2"-long single shot made of cast iron in 1890 by Stevens. $80-100

America, 9 1/8"-long single shot made of cast iron in 1880 by Stevens. $150-125

Army, 5 3/8"-long single shot made of cast iron in 1900 by Stevens. $40-60

Army, 6 1/8"-long single shot made of nickel-plated cast iron in 1916 by Kenton. $40-60

Army, 6 3/4"-long single shot made of nickel-plated cast iron in 1920 by Ideal. $60-80

A Loud Book, 3 1/2" X 5 3/4" single shot made by Kellogg in 1910. $200-250

Bandit, 7 1/4"-long single shot made of nickel-plated cast iron in 1940 by Kenton. $40-50

Bang, 6"-long single shot made of nickel-plated cast iron in 1930 by Kilgore. $30-40

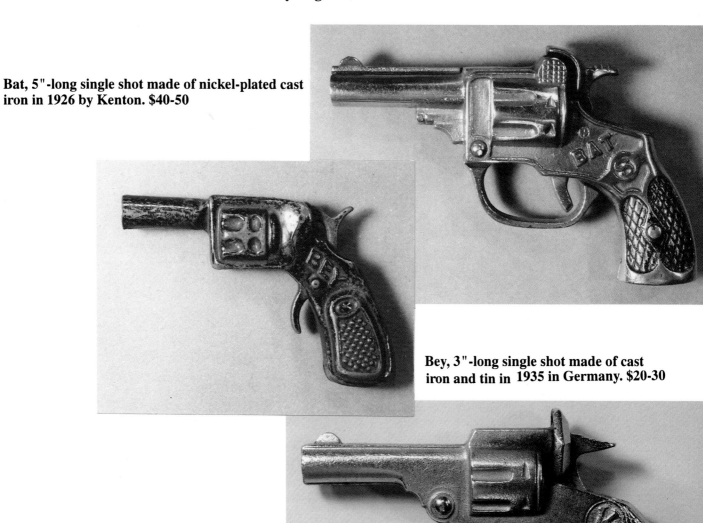

Bat, 5"-long single shot made of nickel-plated cast iron in 1926 by Kenton. $40-50

Bey, 3"-long single shot made of cast iron and tin in 1935 in Germany. $20-30

Big Bill, 5 1/4"-long single shot made of nickel-plated cast iron in 1916 by Kilgore. $40-60

Big Bill, 4 7/8"-long single shot made of nickel-plated cast iron in 1936 by Kilgore. $40-60

Big Boy, 7 1/2"-long single shot made of nickel-plated cast iron in 1922 by Andes. $60-80

Big Chief, 6"-long single shot made of nickel-plated cast iron in 1935 by Kilgore. $30-40

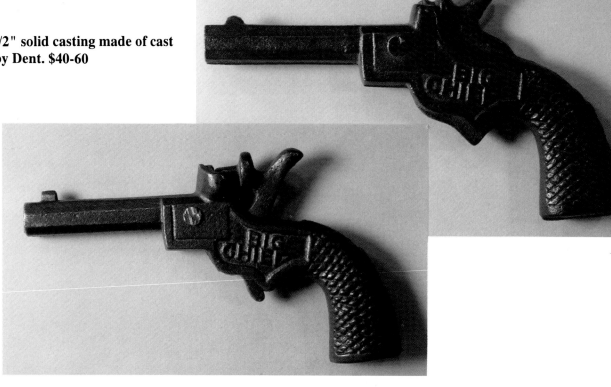

Big Chief, 3 1/2" solid casting made of cast iron in 1930 by Dent. $40-60

Big Chief, 3 1/2"-long single shot made of cast iron in 1930 by Dent. $40-60

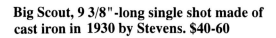

Big Scout, 9 3/8"-long single shot made of cast iron in 1930 by Stevens. $40-60

Big Scout, 7 1/8"-long single shot made of nickel-plated cast iron in 1940 by Stevens. $30-40

Big Smokey, 6 3/4"-long single shot made of nickel-plated cast iron in 1925 by Kilgore. $40-60

Bigger Bang, 5 1/2"-long single shot made of nickel-plated cast iron in 1930 by Kilgore. $40-60

Bigger Bang, 6"-long single shot made of nickel-plated cast iron in 1930 by Hubley. $40-60

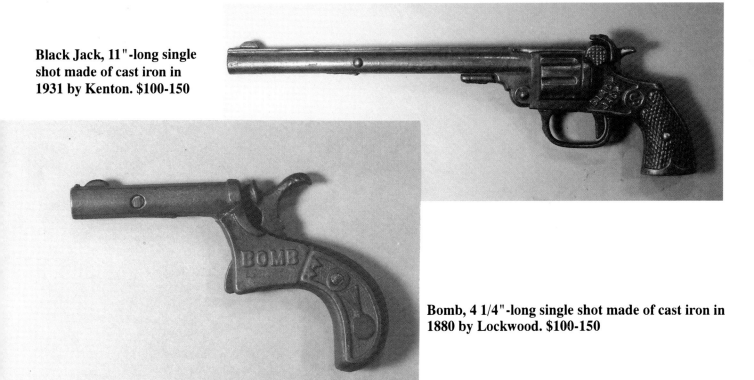

Black Jack, 11"-long single shot made of cast iron in 1931 by Kenton. $100-150

Bomb, 4 1/4"-long single shot made of cast iron in 1880 by Lockwood. $100-150

Boss, 5 1/4"-long single shot made of nickel-plated cast iron in 1900 by Gray Iron. $60-80

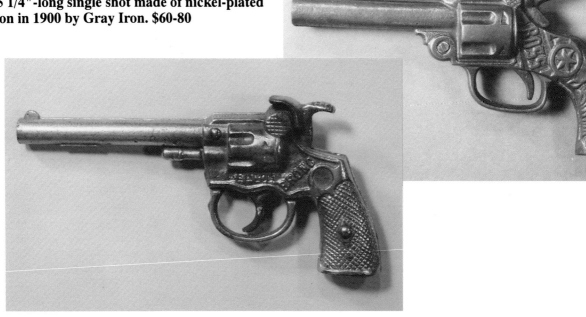

Bronc, 6 1/8"-long single shot made of nickel-plated cast iron in 1937 by Kenton. $40-60

Buc-A-Roo, 8 1/4"-long single shot made of painted cast iron in 1940 by Kilgore. $60-80

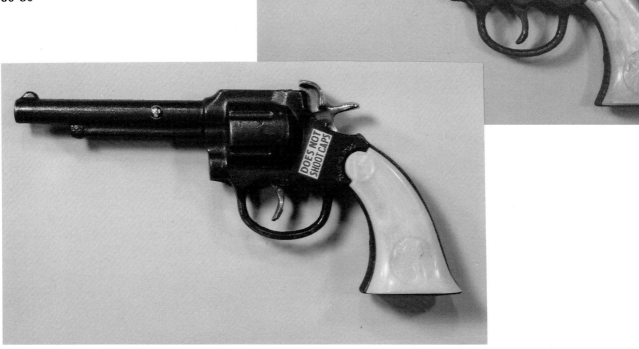

Buc-A-Roo, Dummy 8 1/4"-long single shot made of painted cast iron in 1940 by Kilgore. $60-80

Buck, 3 1/4"-long single shot made of cast iron in 1932 by Hubley. $40-60

Buddy, 4 1/2"-long single shot made of nickel-plated cast iron in 1930 by Dent. $40-60

Buddy, 6 1/4"-long single shot made of nickel-plated cast iron in 1935 by Stevens. $40-60

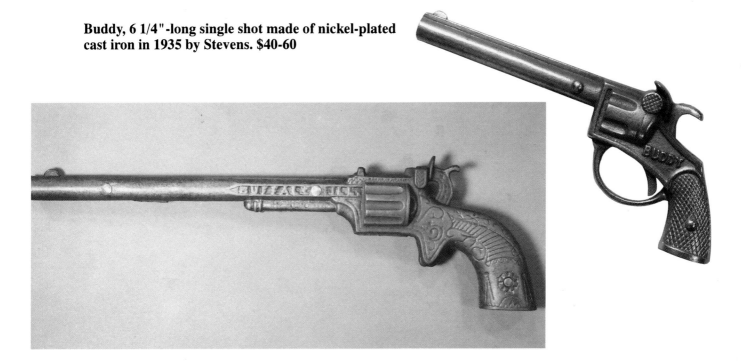

Buffalo Bill, 11 3/4 " long single shot made of cast iron in 1890 by Stevens. $250-325

Buffalo Bill, 11 3/8 " long single shot made of cast iron in 1923 by Stevens. $150-250

Bull Dog, 6"-long single shot made of nickel-plated cast iron in 1935 by Hubley. $40-60

Bull Dog, 6"-long single shot made of painted cast iron in 1935 by Hubley. $40-60

Bull, 4 3/8"-long single shot made of cast iron in 1930 by Hubley. $30-40

Bunker Hill, 5 3/4"-long single shot made of nickel-plated cast iron in 1925 by National. $40-60

Buster, 5 1/8"-long single shot made of nickel-plated cast iron in 1930 by Dent. $30-40

Caddy, 5 1/2"-long single shot made of nickel-plated cast iron in 1930 by Hubley. $30-40

Cadet, 5 3/8"-long single shot made of nickel-plated cast iron in 1930 by Hubley. $40-60

Cal, 5 1/2"-long single shot made of nickel-plated cast iron in 1925 by Stevens. $40-60

Cavalry, 9"-long single shot made of nickel-plated cast iron made by Autor. $150-225

Chief, 6 1/8"-long single shot made of nickel-plated cast iron in 1930 by Hubley. $40-60

Chief, Chief on the other side 5 1/8"-long single shot made of cast iron in 1935 by Kilgore. $40-60

Chieftain, 11"-long single shot made of nickel-plated cast iron in 1920 by National. $150-200

Clip Jr., 5 3/4"-long single shot made of nickel-plated cast iron in 1935 by Stevens. $30-40

Clip, 6 1/4"-long single shot made of nickel-plated cast iron in 1935 by Hubley. $30-40

Colt, 5 1/2"-long single shot made of cast iron in 1907 by Stevens. $60-80

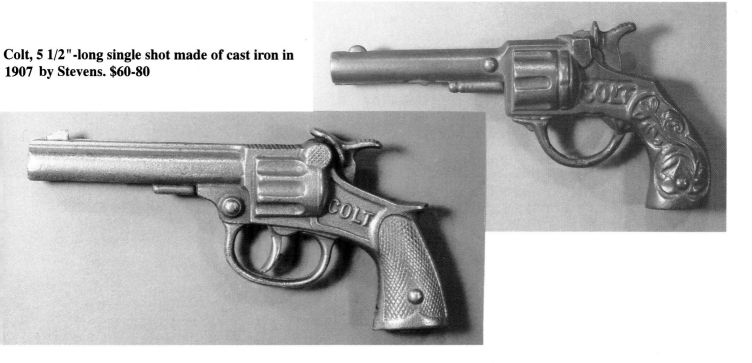

Colt, 6 1/2"-long single shot made of nickel-plated cast iron in 1935 by Stevens. $40-60

Columbia, 8 3/4" long single shot made of cast iron in 1890 by Stevens. $250-350

Comet, 5 1/2"-long single shot made of cast iron in 1885 by Stevens. $250-300

Comet, 7"-long single shot made of cast iron. $60-80

Comet, 7 1/8"-long single shot made of nickel-plated cast iron in 1925 by Stevens. $60-80

Cop, 5"-long single shot made of nickel-plated cast iron in 1930 by Hubley. $40-60

Cowboy, 12"-long single shot made of nickel-plated cast Iron in 1930 by Stevens. $250-300

Cowboy, 3 1/2"-long single shot made of cast iron in 1935 by Stevens. $30-40

Crack, 5"-long single shot made of nickel-plated cast iron in 1925 by Stevens. $40-60

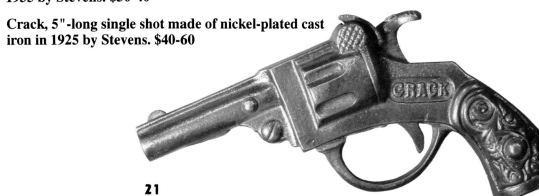

Crack, 5"-long single shot made of nickel-plated cast iron in 1925 by Stevens. $40-60

Cub, 4"-long single shot made of nickel-plated cast iron in 1930 by Kenton. $40-60

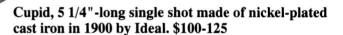

Cupid, 5 1/4"-long single shot made of nickel-plated cast iron in 1900 by Ideal. $100-125

D - Lux, 9"-long single shot made of nickel-plated cast iron in 1935 by Kilgore. Has broken trigger guard. $60-80

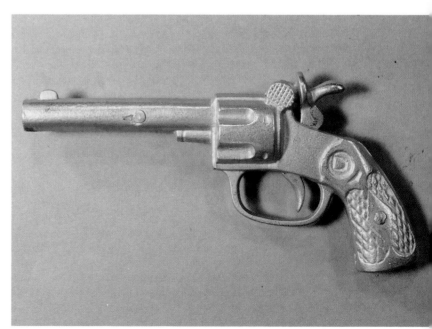

D, 5 1/8"-long single shot made of nickel-plated cast iron in 1930 by Dent. $30-40

Daisy, 4 3/4"-long single shot made of cast iron in 1875 by Stevens. $150-250

Daisy, 4 1/8"-long single shot made of nickel-plated cast iron in 1925 by Hubley. $30-40

Daisy, 4 1/8" long single shot made of nickel-plated cast iron in 1935. $30-40

Darb, 5 1/2"-long single shot made of nickel-plated cast iron in 1926 by Kenton. $30-40

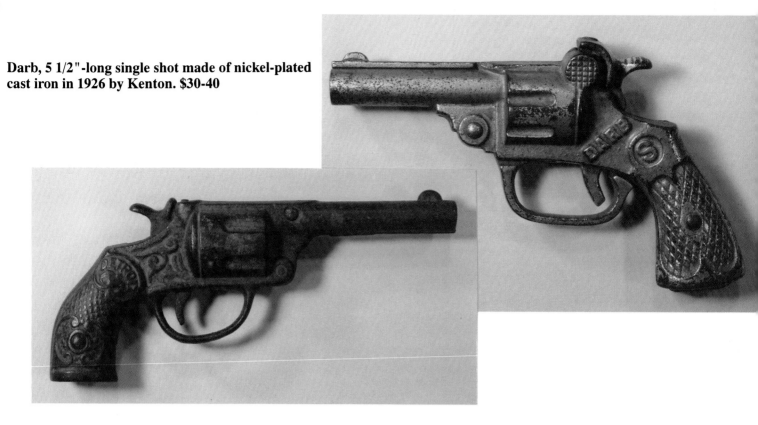

Detroit, 6 5/8"-long single shot made of cast iron in 1910. $100-150

Dik, 4 3/4"-long single shot made of nickel-plated cast iron in 1930 by Kenton. Note DIK is straight. $40-60

Dick, 6" long, single shot made of cast iron in 1940 by Hubley. $30-40

Dix, 4 5/8"-long single shot made of cast iron in 1926 by Kenton. $40-60

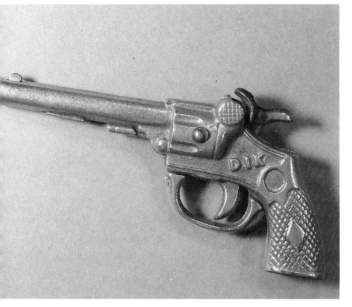
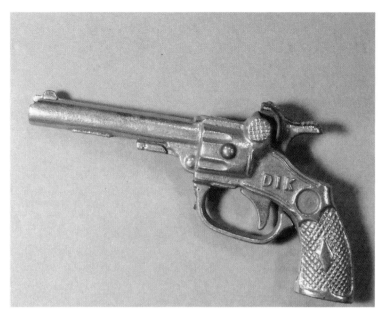

Dik, 4 3/4"-long single shot made of nickel-plated cast iron in 1940 by Kenton. $30-40

Dik, dummy 4 3/4"-long single shot made of nickel-plated cast iron in 1940 by Kenton. $40-60

Doc, Pat Sep. 11 1923 4 1/2"-long single shot made of nickel-plated cast iron in 1926 by Kenton. $60-80

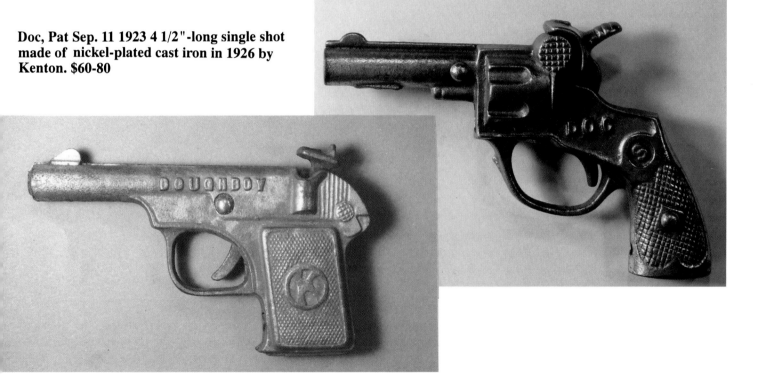

Doughboy, 5"-long single shot made of nickel-plated cast iron in 1920 by Kilgore. $60-80

Doughboy, 5"-long single shot made of nickel-plated cast iron in 1920 by Kilgore. $60-80

Dude, 5 3/4"-long single shot made of cast iron in 1941 by Kenton. $50-70

Eagle, 7 1/4"-long single shot made of nickel-plated cast iron in 1895 by Stevens. $125-150

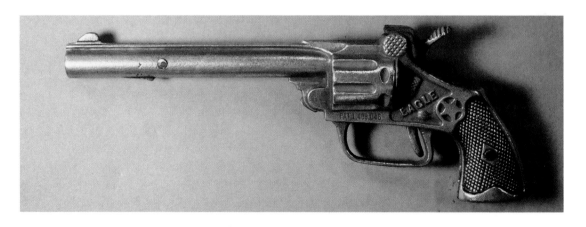

Eagle, 8 1/2"-long single shot made of nickel-plated cast iron in 1935 by Hubley. $125-150

Eber, 5"-long single shot made of painted cast iron in Germany. $20-40

Echo, 4 1/2"-long single shot made of nickel-plated cast iron in 1930 by Stevens. $30-40

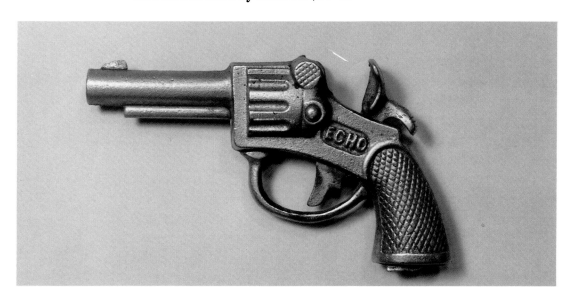

Echo, dummy 4 1/2"-long single shot made of nickel-plated cast iron in 1930 by Stevens. $60-80

44 Cal, 7 1/2"-long single shot made of cast iron in 1930 by Kenton. $60-80

45, 5 1/8"-long single shot made of cast iron in 1910 by Stevens. $60-80

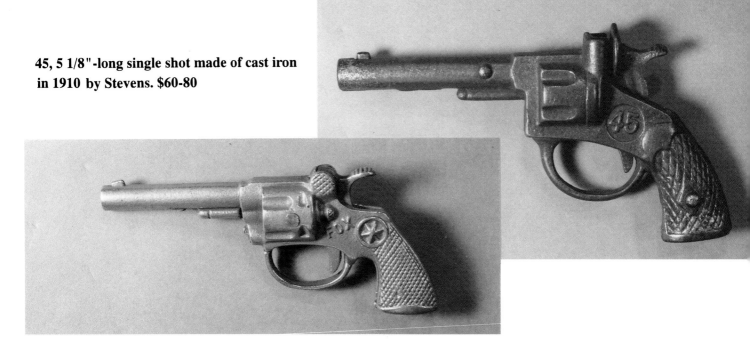

Fox, 4 1/2"-long single shot made of nickel-plated cast iron in 1935 by Hubley. $30-40

Gip, 5 1/4"-long single shot made of cast iron in 1900 by Stevens. $60-80

Gem, 3 3/4"-long single shot made of nickel-plated cast iron in 1924 by Stevens. $30-40

Hero, 5 1/4 " long single shot made of nickel-plated cast iron in 1937 by Stevens. $30-40

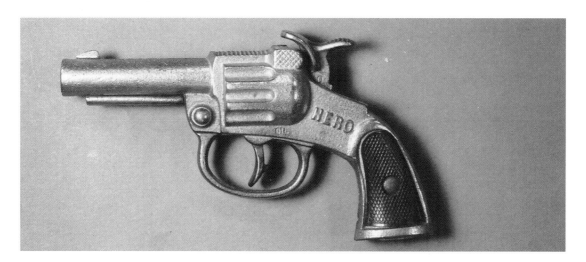

Hero, dummy 5 1/4"-long single shot made of nickel-plated cast iron in 1935 by Stevens. $40-60

Hero, dummy 5 1/4 " long single shot made of nickel-plated cast iron in 1937 by Stevens. $40-60

Hi - Ho, dummy 7"-long single shot made of nickel-plated cast iron in 1940 by Stevens. $40-60

Hio, 5 1/8"-long single shot made of nickel-plated cast iron in 1926 by Kenton. $30-40

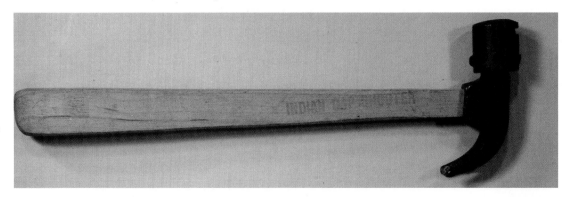

Indian Cap Hammer, 10 3/4"-long made of cast iron in 1890. $100-150

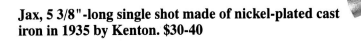

Jax, 5 3/8"-long single shot made of nickel-plated cast iron in 1935 by Kenton. $30-40

Jr. Ranger 32 Cal, Jr. Ranger on other side. 6 1/8"-long single shot made of cast iron in 1935 by Mordt. $40-60

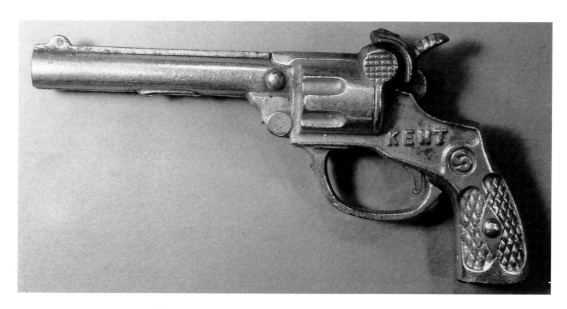

Kent, 5 3/4 " long single shot made of nickel-plated cast iron in 1932 by Kenton. $30-40

Kido, 5 3/8"-long single shot made of nickel-plated cast iron in 1937 by Kenton. $40-60

Liberty, 7 1/8"-long single shot made of cast iron in 1875 by Ives. $200-300

Lion, Pat. June 17 1890 5 1/4"-long single shot made of cast iron in 1890 by Stevens. $300-400

Lion, 3 1/4" long single shot made of nickel-plated cast iron in 1920 by Hubley. $60-80

Little Bill, 5"-long single shot made of cast iron in 1941 by Kilgore. $30-40

Long Boy, 11"-long single shot made of nickel-plated cast iron in 1922 by Kilgore. $100-150

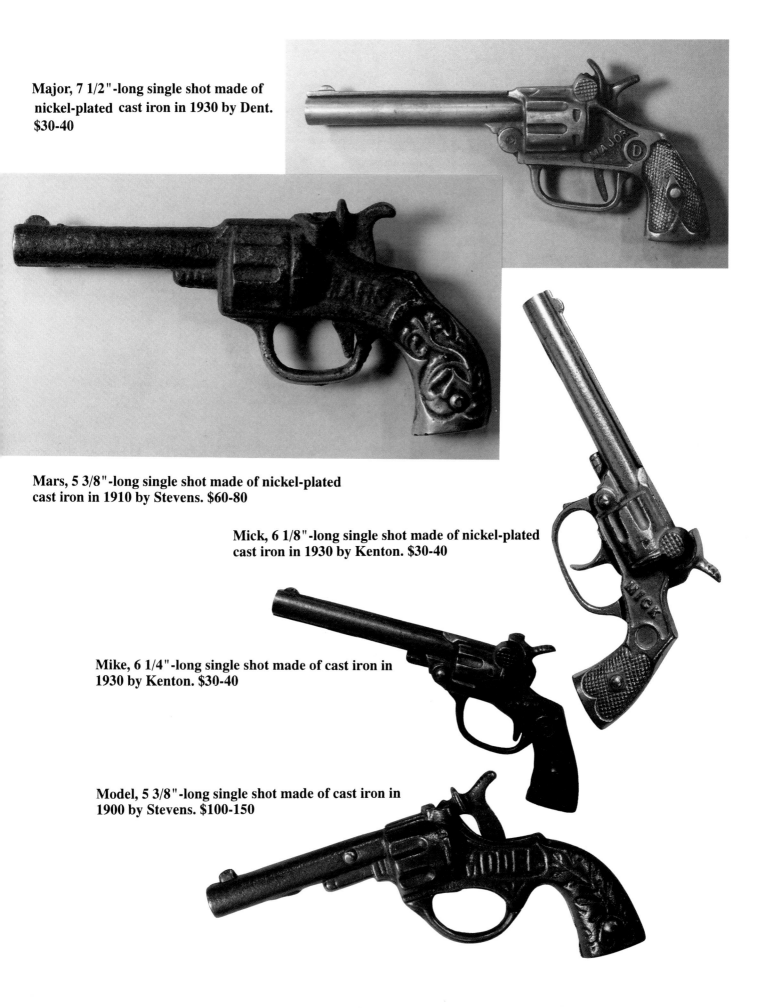

Major, 7 1/2"-long single shot made of nickel-plated cast iron in 1930 by Dent. $30-40

Mars, 5 3/8"-long single shot made of nickel-plated cast iron in 1910 by Stevens. $60-80

Mick, 6 1/8"-long single shot made of nickel-plated cast iron in 1930 by Kenton. $30-40

Mike, 6 1/4"-long single shot made of cast iron in 1930 by Kenton. $30-40

Model, 5 3/8"-long single shot made of cast iron in 1900 by Stevens. $100-150

Mohawk, 8 3/8"-long single shot made of cast iron in 1930 by Hubley. $60-80

Mohican, 6 1/4"-long single shot made of nickel-plated cast iron in 1930 by Dent. $80-100

Mordt, 8"-long single shot made of cast iron in 1932 by Mordt. $100-150

Mut, 4 3/8"-long single shot made of nickel-plated cast iron in 1925 by Kenton. $30-40

Nat, 4 5/8"-long single shot made of nickel-plated cast iron in 1930 by Stevens. $30-40

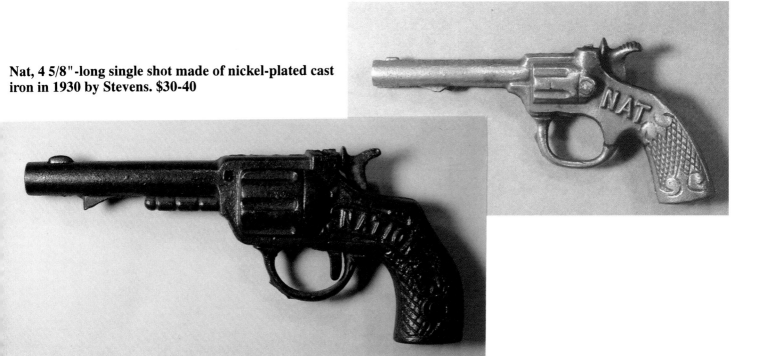

National, 6 5/8"-long single shot made of cast iron in 1920 by Stevens. $60-80

Navy, 5 3/4"-long single shot made of nickel-plated cast iron in 1907 by Ideal. Note rivet at bottom of the grip. $125-150

Navy, 5 3/4"-long single shot made of nickel-plated cast iron in 1907 by Ideal. Note rivet in center of the grip. $125-150

Navy, 5 3/8"-long single shot made of nickel-plated cast iron in 1923 by Kenton. $40-60

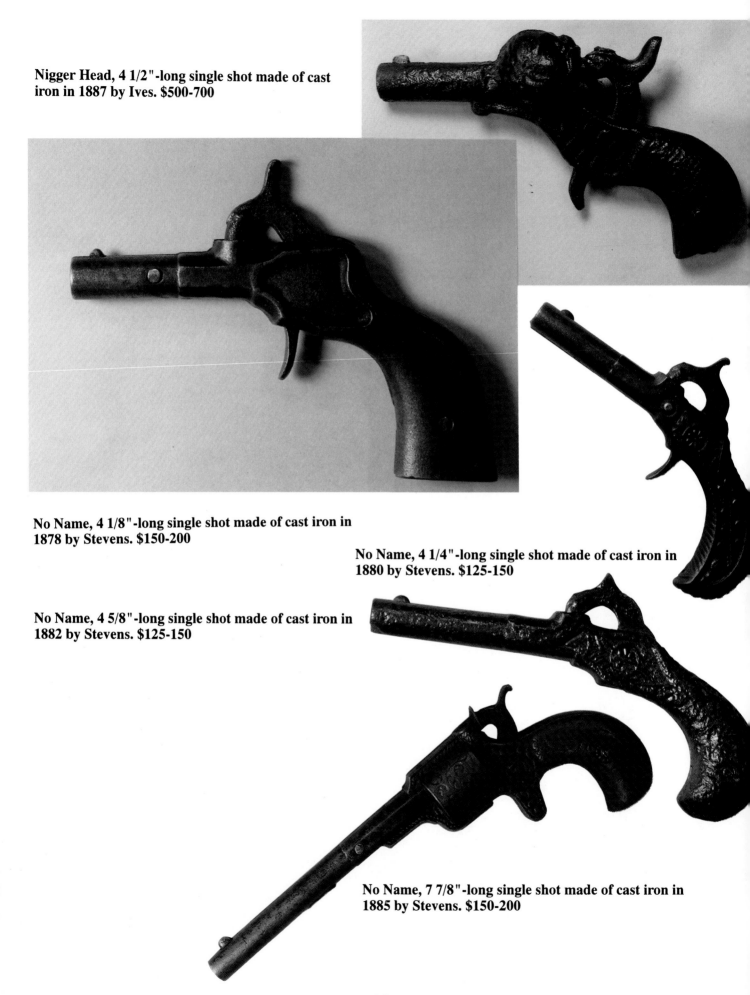

Nigger Head, 4 1/2"-long single shot made of cast iron in 1887 by Ives. $500-700

No Name, 4 1/8"-long single shot made of cast iron in 1878 by Stevens. $150-200

No Name, 4 1/4"-long single shot made of cast iron in 1880 by Stevens. $125-150

No Name, 4 5/8"-long single shot made of cast iron in 1882 by Stevens. $125-150

No Name, 7 7/8"-long single shot made of cast iron in 1885 by Stevens. $150-200

No Name, 4 3/4"-long single shot made of cast iron in 1885 by Stevens. $150-200

No Name, 7"-long single shot made of cast iron in 1895 by Stevens. $125-150

No Name, 2 7/8"-long single shot made of nickel-plated cast iron in 1900. $80-100

No Name, 6 5/8"-long single shot made of nickel-plated cast iron in 1904 by Kenton. $80-100

No Name, 6 3/4"-long single shot made of cast iron in 1910 by Stevens. $80-100

No Name, 5 1/4"-long single shot made of cast iron in 1911 by Kenton. $80-100

No Name, 11 1/2"-long single shot made of cast iron in 1915 by Kenton. $150-200

No Name, 3 1/2"-long single shot made of nickel-plated cast iron in 1920 by Hubley. $30-40

No Name, 4 3/8"-long single shot made of nickel-plated cast iron in 1920 by Kenton. $40-60

No Name, 3 1/2"-long single shot made of cast iron in 1920 by Kenton. $40-60

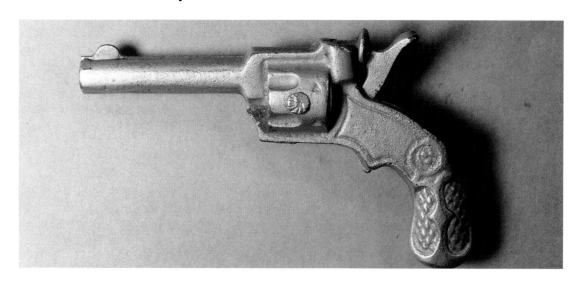

No Name, 3 3/4"-long single shot made of nickel-plated cast iron in 1924 by Kenton. $40-60

No Name, 6 1/2", made of cast iron in 1925 by National. $40-60

No Name, 4 7/8"-long single shot made of cast iron in 1932 by Ives. $80-100

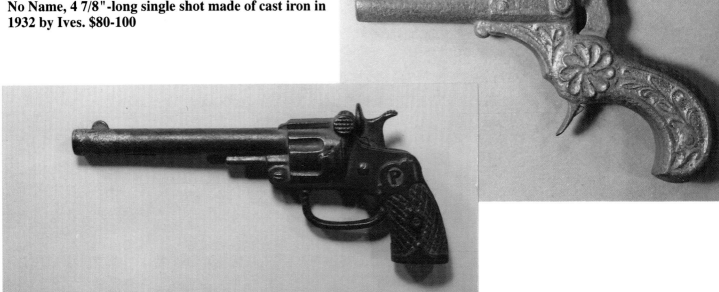

No Name, 6 1/2"-long single shot made of cast iron in 1930 by Kilgore. $30-40

No Name, 5 1/2"-long single shot made of nickel-plated cast iron in 1935 by Kenton. $30-40

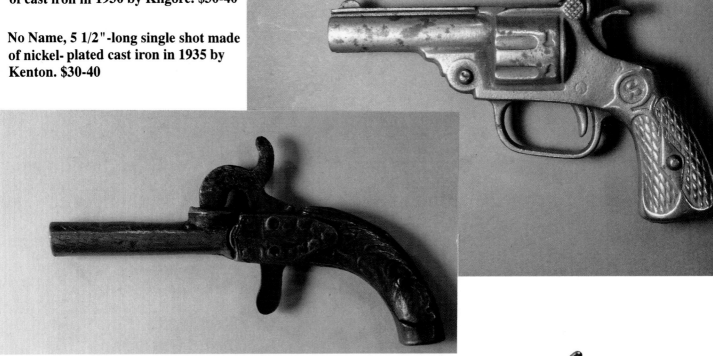

No Name, 3 1/4"-long single shot made of cast iron. $100-150

No Name, 4 1/2"-long single shot made of tin. $20-30

Novelty, 5"-long single shot made of cast iron in 1885 by Lockwood. $150-200

101 Ranch, 11 1/2"-long single shot made of nickel-plated cast iron in 1930 by Hubley. $150-175

1776 - 1876, 5 1/4"-long single shot made of cast iron in 1876 by Stevens. $250-350

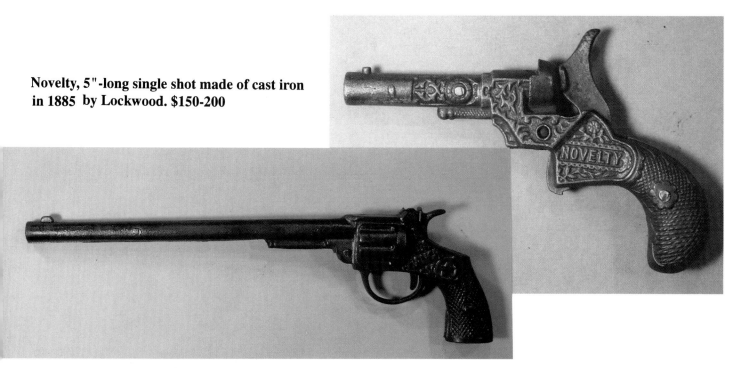
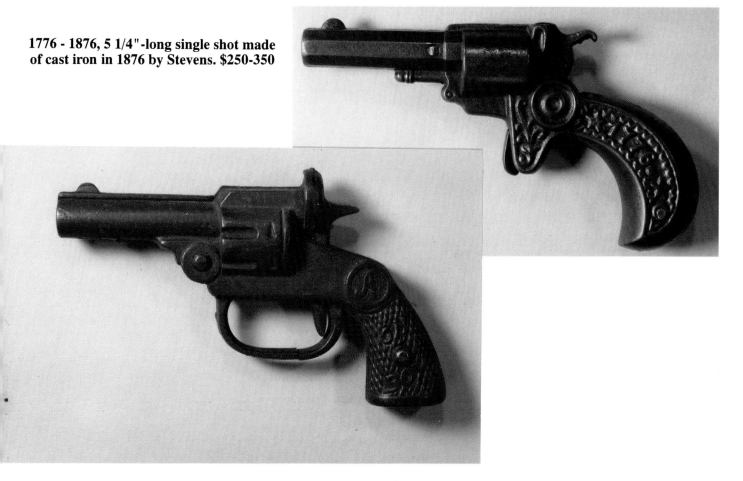

Oh Boy, 5 1/2"-long single shot made of cast iron in 1922 by Andes. $30-40

Ok, Pat June 17 1890, 3 1/4"-long single shot made of cast iron in 1890 by Stevens. $80-100

Oke, 6"-long single shot made of cast iron in 1930 by Kilgore. $30-40

Old Ironsides, 10 1/2"-long single shot made of cast iron in 1922 by National. $150-175

P, 3 3/4"-long single shot made of cast iron in 1930 by Kilgore. $30-40

P, 9 1/2"-long single shot made of cast iron in 1930 by Kilgore. $100-125

Pat, 6 1/8"-long single shot made of nickel-plated cast iron in 1931 by Kenton. $30-40

Peach, 5 1/4"-long single shot made of nickel-plated cast iron in 1905. $250-300

Pet, 4 1/8"-long single shot made of cast iron in 1924 by Stevens. $30-40

Pet, 4 5/8"-long single shot made of nickel-plated cast iron in 1934 by Hubley. $30-40

Pluck, 3 1/2"-long single shot made of cast iron in 1925 by Stevens. $30-40

Pluck, 3 3/8"-long single shot made of cast iron in 1930 by Stevens. Note hammer variation. $30-40

Pono, 5 1/8"-long single shot made of nickel-plated cast iron in 1936 by Kenton. $30-40

Pony, 3 3/4"-long single shot made of nickel-plated cast iron in 1890 by Ives. $125-175

Pony, 5 1/8"-long single shot made of nickel-plated cast iron in 1930 by Kenton. Note ridge around name. $30-40

Pony, 5 1/8"-long single shot made of cast iron in 1930 by Kenton. $30-40

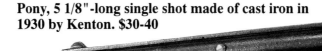

President, President is on the other side shoots caps and rubber bands 8 3/4"-long single shot made of cast iron in 1925 by Kilgore. $100-150

Pup, 3 3/4"-long single shot made of nickel-plated cast iron in 1930 by Kenton. $30-40

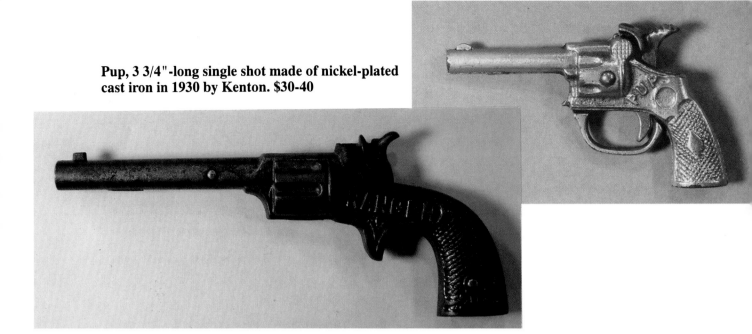

Ranger, 7 3/8"-long single shot made of cast iron in 1890 by Stevens. $125-150

Ranger, 6 3/4"-long single shot made of cast iron in 1923 by Kenton. $40-60

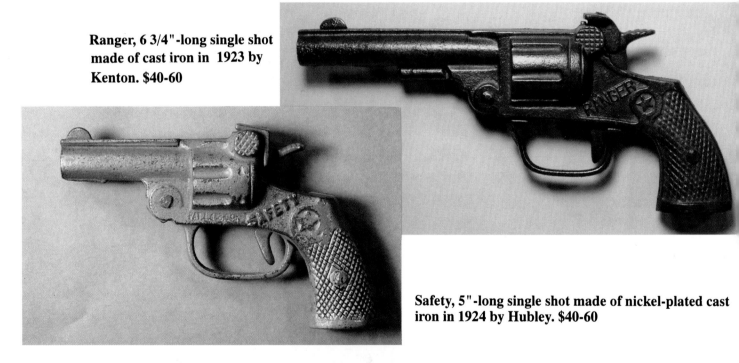

Safety, 5"-long single shot made of nickel-plated cast iron in 1924 by Hubley. $40-60

Rodeo, 11 1/4"-long single shot made of cast iron in 1924 by Hubley. Note 3" of this guns barrel is broken off. $60-80

Savage, 7"-long single shot made of nickel-plated cast iron in 1925 by Hubley. $40-60

Scout, 7"-long single shot made of nickel-plated cast iron in 1890 by Stevens. $100-125

Scout, 6 3/4"-long single shot made of nickel-plated cast iron in 1935 by Stevens. $40-60

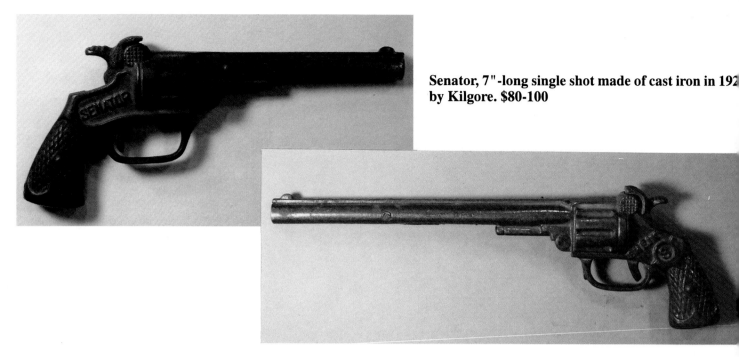

Senator, 7"-long single shot made of cast iron in 192[?] by Kilgore. $80-100

Sheik, 10 1/2"-long single shot made of cast iron in 1929 by Kenton. $125-150

Slick, 6 1/4"-long single shot made of nickel-plated cast iron in 1930 by Kenton. $40-60

Sliko, 6 1/2"-long single shot made of cast iron in 1930 by Kenton. $40-60

Sport, 7 1/2"-long single shot made of cast iron in 1930 by Kilgore. $40-60

Spy, 4 1/4"-long single shot made of nickel-plated cast iron in 1936 by Kilgore. $30-40

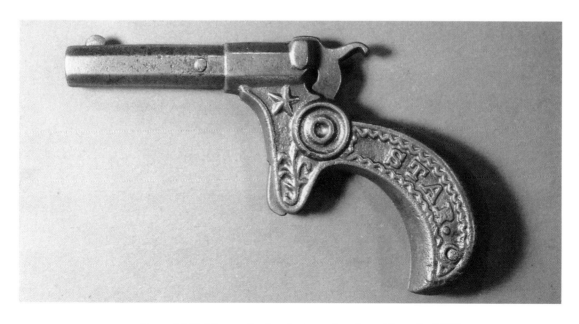

Star, 4 3/8" long single shot made of cast iron in 1882 by Stevens. $200-300

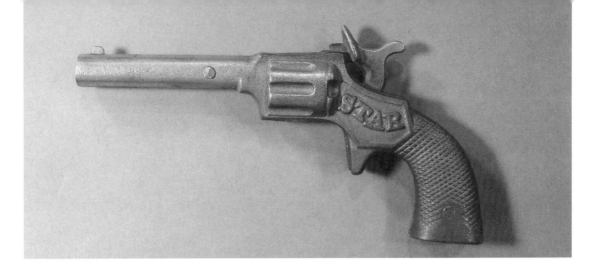

Star, 6 7/8"-long single shot made of nickel-plated cast iron in 1890 by Stevens. $150-200

Star, 6 1/4"-long single shot made of nickel-plated cast iron in 1910 by Stevens. $60-80

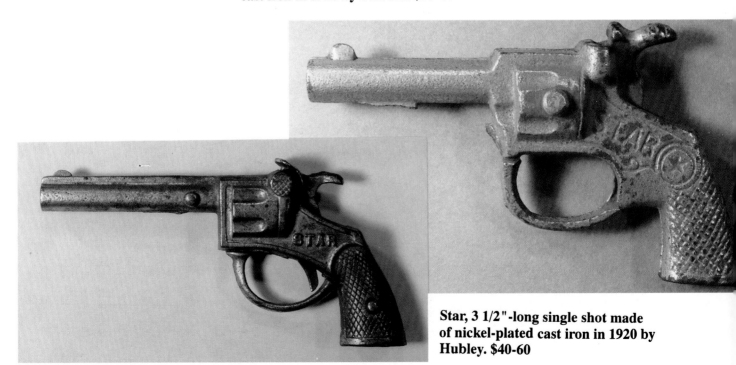

Star, 3 1/2"-long single shot made of nickel-plated cast iron in 1920 by Hubley. $40-60

Star, 5 1/4"-long single shot made of nickel-plated cast iron in 1930 by Stevens. $40-60

Star, double shot 6 3/4"-long made of nickel-plated cast iron. $100-150

32, 4 3/4"-long single shot made of nickel-plated tin. $60-80

Super, 8 3/4"-long single shot made of nickel-plated cast iron in 1930 by Kenton. $60-80

Target, 8"-long single shot made of cast iron in 1935 by Hubley. $40-60

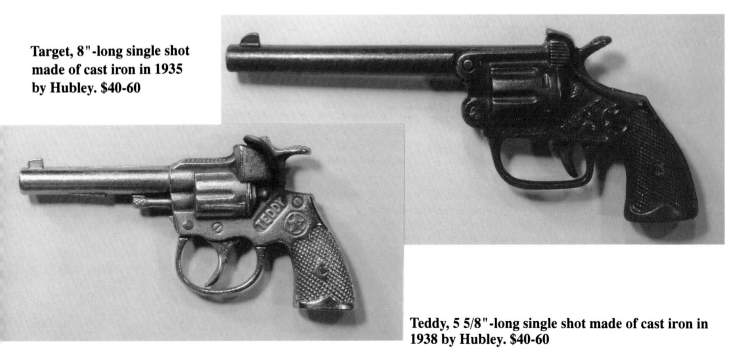

Teddy, 5 5/8"-long single shot made of cast iron in 1938 by Hubley. $40-60

Texas, 6 1/2"-long single shot made of cast iron in 1930 by Kenton. $40-60

Tiger, 6 7/8"-long single shot made of cast iron in 1935 by Hubley. $40-60

Trooper, 7 1/8"-long single shot made of cast iron in 1938 by Hubley. $40-60

Two Time, shoots caps and rubber bands 9 1/4"-long single shot made of nickel-plated cast iron in 1929 by Kenton. $150-175

Victor, 5 3/4"-long single shot made of cast iron in 1880 by Stevens. $150-200

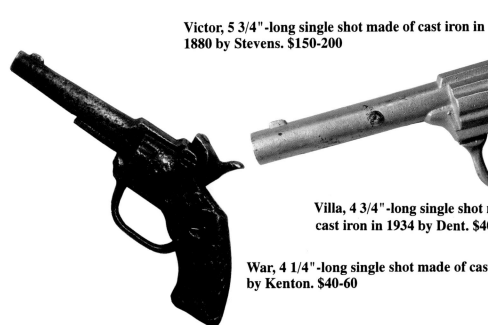

Villa, 4 3/4"-long single shot made of nickel-plated cast iron in 1934 by Dent. $40-60

War, 4 1/4"-long single shot made of cast iron in 1923 by Kenton. $40-60

Western, 7"-long single shot made of cast iron in 1931 by Kenton. $80-100

Western, 7 1/4"-long single shot made of nickel-plated cast iron in 1939 by Kenton. $60-80

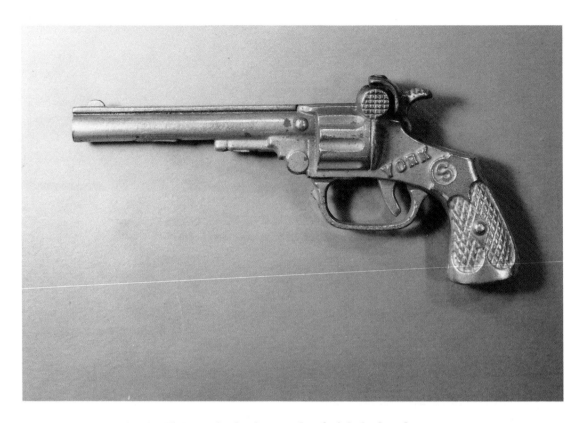

York, 7"-long single shot made of nickel-plated cast iron in 1930 by Kenton. $60-80

Zip, 5"-long single shot made of cast iron in 1935 by Hubley. $40-60

CAST IRON AUTOMATIC CAP GUNS

Army 45, 7"-long automatic made of cast iron in 1940 by Hubley. $100-150

Army 45, 7"-long automatic made of nickel-plated cast iron in 1940 by Hubley. $125-175

Biff Jr., 4 1/8"-long automatic made of cast iron in 1937 by Kenton. $40-60

Bang O, 7 1/2" automatic made of nickel-plated cast iron in 1940 by Stevens. $80-100

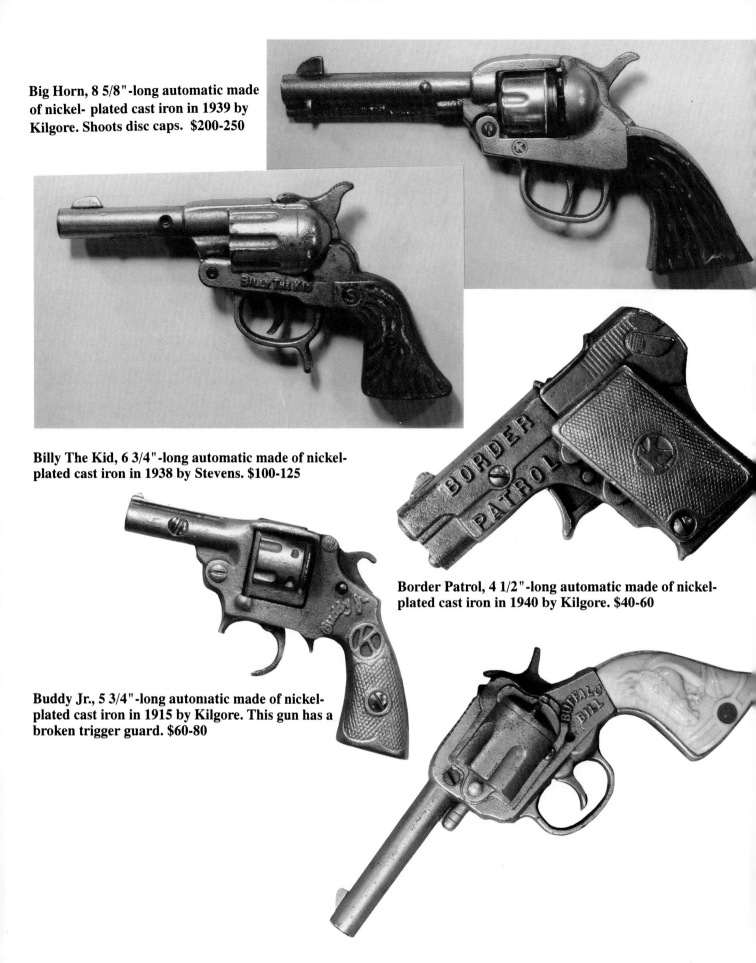

Big Horn, 8 5/8"-long automatic made of nickel- plated cast iron in 1939 by Kilgore. Shoots disc caps. $200-250

Billy The Kid, 6 3/4"-long automatic made of nickel-plated cast iron in 1938 by Stevens. $100-125

Border Patrol, 4 1/2"-long automatic made of nickel-plated cast iron in 1940 by Kilgore. $40-60

Buddy Jr., 5 3/4"-long automatic made of nickel-plated cast iron in 1915 by Kilgore. This gun has a broken trigger guard. $60-80

Buffalo Bill, 7 3/4"-long automatic made of nickel-plated cast iron in 1940 by Stevens. $80-100

Bulls Eye, 6 1/2" automatic made of nickel-plated cast iron in 1940 by Kilgore. $80-100

Captain, 4 1/4"-long automatic made of nickel-plated cast iron in 1940 by Kilgore. $40-60

Bulls Eye, 6 1/2"-long automatic made of cast iron in 1940 by Kenton. $80-100

Captain, 4 1/4"-long automatic made of painted cast iron in 1940 by Kilgore. $40-60

Champ, 5"-long automatic made of nickel-plated cast iron in 1940 by Hubley. $60-80

Buster, Pat. July 2 1907 5 1/4"-long automatic made of nickel-plated cast iron in 1910 by Kilgore. $100-125

Clipper, 4 1/8"-long automatic made of nickel-plated cast iron in 1935 by Kilgore. $40-60

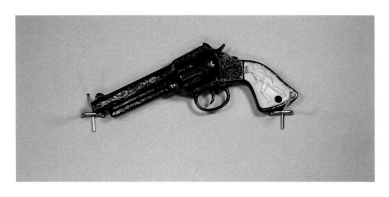

Cowboy King, 9" automatic made of nickel-plated cast iron in 1940 by Stevens. $125-150

Cowboy, 8"-long automatic made of nickel-plated cast iron in 1940 by Hubley. $80-100

Dandy, 5 3/4"-long automatic made of nickel-plated cast iron in 1937 by Hubley. $60-80

Derby, 7"-long automatic made of nickel-plated cast iron in 1930 by Hubley. $40-60

Dixie 7"-long automatic made of cast iron in 1937 by Kenton. $60-80

49-ER, 9"-long automatic made of nickel-plated cast iron in 1940 by Stevens. $150-200

Federal Repeater No. 1, 5 3/8"-long automatic made of nickel-plated cast iron in 1915 by Federal. $60-80

Flash, 6 1/4"-long automatic made of cast iron in 1934 by Hubley. $40-60

G-Man, 6"-long automatic made of painted cast iron. Caps are put in a clip. Made by Kilgore in 1935. $80-100

G-Man, 6"-long automatic made of nickel-plated cast iron. Caps are put in the clip. Made by Kilgore in 1935. $100-125

Gat, 8 1/4" automatic made of nickel-plated cast iron in 1935 by Hubley. $80-100

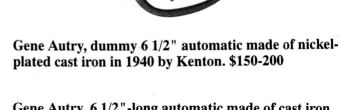

Gene Autry, dummy 6 1/2" automatic made of nickel-plated cast iron in 1940 by Kenton. $150-200

Gene Autry, 6 1/2"-long automatic made of cast iron in 1940 by Kenton. $125-150

Gene Autry, 6 1/2" automatic made of cast iron in 1940 by Kenton. $125-150

Gene Autry, 6 1/2"-long automatic made of nickel- plated cast iron in 1940 by Stevens. $125-150

Gene Autry, dummy 6 1/2"-long automatic made of nickel-plated cast iron in 1940 by Kenton. $150-175

Gene Autry, 7 3/4"-long automatic made of cast iron in 1940 by Stevens. $150-175

Gene Autry, 8 3/8" automatic made of cast iron in 1939 by Kenton. $150-175

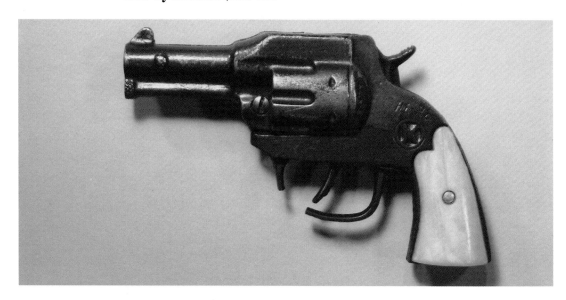

Hi-Ho, 6 1/2"-long automatic made of cast iron in 1940 by Kilgore. Note this gun has a broken trigger guard. $40-60

Hi-Ho, horse on grip 6 1/2"-long automatic made of cast iron in 1940 by Kilgore. $40-60

Hi-Ho, 6 1/2"-long automatic made of cast iron in 1940 by Kilgore. $40-60

Hi Ranger, 7 3/4"-long automatic made of nickel-plated cast iron in 1940 by Stevens. $60-80

Invincible 50 Shot, 5 1/2"-long automatic made of painted cast iron in 1915 by Kilgore. $40-60

Invincible, 5 1/4" automatic made of nickel-plated cast iron in 1935 by Kilgore. $40-60

Invincible, 6 1/2"-long automatic made of cast iron in 1930 by Kilgore. $40-60

Jr. Police Chief, 3 7/8"-long automatic made of nickel-plated cast iron in 1941 by Kenton. $40-60

Kilgore, 5 1/4"-long automatic made of nickel-plated cast iron in 1912 by Federal. $40-60

Law Maker, 8 3/8"-long automatic made of painted cast iron in 1941 by Kenton. $125-150

Lawmaker, 8 3/8" automatic made of cast iron in 1941 by Kenton. This gun has a broken hammer. $125-150

Lone Ranger, 8 1/2"-long automatic made of cast iron in 1938 by Kilgore. $150-200

Lone Ranger, 8 1/2"-long automatic made of nickel-plated cast iron in 1940 by Kilgore. $175-225

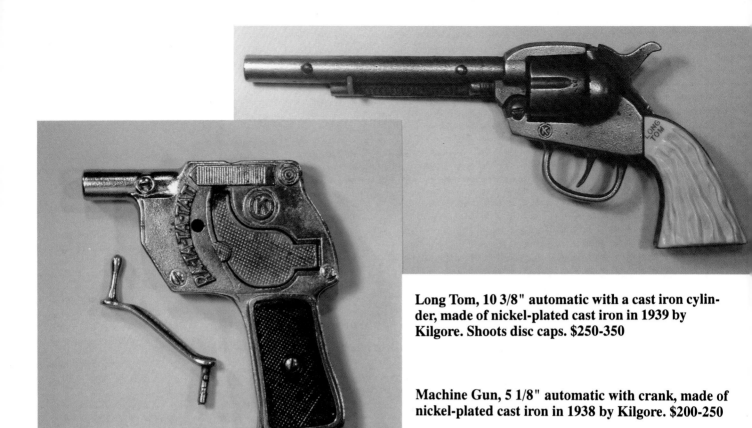

Long Tom, 10 3/8" automatic with a cast iron cylinder, made of nickel-plated cast iron in 1939 by Kilgore. Shoots disc caps. $250-350

Machine Gun, 5 1/8" automatic with crank, made of nickel-plated cast iron in 1938 by Kilgore. $200-250

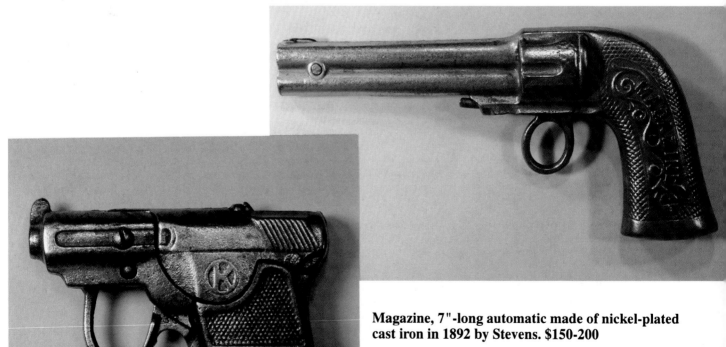

Magazine, 7"-long automatic made of nickel-plated cast iron in 1892 by Stevens. $150-200

Mascot, 4"-long automatic made of nickel-plated cast iron in 1936 by Kilgore. $40-60

Master, 4 5/8"-long automatic made of nickel-plated cast iron in 1922 by Andas. $60-80

National, 4 7/8"-long automatic made of cast iron in 1909 by National. $80-100

National, 5"-long automatic made of nickel-plated cast iron in 1911 by National. $80-100

National, 3 3/4"-long automatic made of nickel-plated cast iron in 1915 by National. $60-80

National, 3 5/8"-long automatic made of nickel-plated cast iron in 1920 by National. $40-60

National, 4 1/4" automatic made of cast iron in 1925 by National. $40-60

National No. 380, 7"-long automatic made of nickel-plated cast iron in 1930 by National. $40-60

Oh Boy, Made in the USA; PAT'D AUG 8, 1933 4 1/8" automatic made of cast iron in 1933 by Kilgore. $125-150

Oh Boy, 4 1/8" automatic made of cast iron in 1933 by Kilgore. $125-150

Ok, 3 3/4" automatic made of cast iron in 1935. $40-60

Pal, 4"-long automatic made of cast iron in 1930 b Kilgore. $40-60

Police Chief, 4"-long automatic made of nickel-plated cast iron in 1938 by Kenton. $40-60

Patrol, 6"-long automatic made of nickel-plated cast iron in 1939 by Kilgore. $40-60

Police, 5 1/4"-long automatic made of painted cast iron in 1945 by Kilgore. $60-80

Peace Maker, 9" automatic made of nickel-plated cast iron in 1940 by Stevens. $150-200

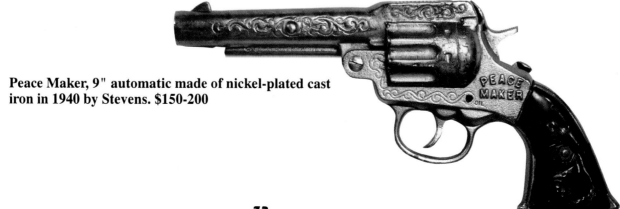

Premiere Safety, 6 1/4"-long automatic made of nickel-plated cast iron in 1914 by Kilgore. $40-60

Presto, 5 1/8"-long automatic made of nickel-plated cast iron in 1940 by Kilgore. $40-60

Ranger, 8 1/2"-long automatic made of nickel-plated cast iron in 1940 by Kilgore. $100-150

Ranger, 7 3/8"-long automatic made of cast iron in 1940 by Stevens. $80-100

Rex, 3 7/8"-long automatic made of painted cast iron in 1939 by Kilgore. $40-60

Ranger, 5 3/8"-long automatic made of nickel-plated cast iron in 1920 by Kilgore. $40-60

Riot Gun, 5 3/8" automatic with crank, made of nickel-plated cast iron in 1932 by Kenton. $150-200

Rex, 3 7/8"-long automatic made of nickel-plated cast iron in 1939 by Kilgore. $40-60

Six Shot, 6 3/4"-long automatic made of cast iron in 1895 by Stevens. $200-150

Safety 50 Shot, 6 1/4"-long automatic made of cast iron in 1925 by Stevens. $40-60

Safety First, 3 3/8"-long automatic made of nickel-plated cast iron in 1925 by Victory Spark. $80-100

Sharpshooter, 6 1/4"-long automatic made of cast iron in 1935 by Kilgore. $80-100

Six Shooter, 6 7/8"-long automatic made of cast iron in 1935 by Kilgore. Shoots disc caps. $40-60

Six Shooter Automatic, 6 1/2"-long automatic made of cast iron in 1934 by Kilgore. Shoots disc caps. $40-60

Six Shooter, 7"-long automatic made of nickel-plated cast iron in 1935 by Kilgore. Shoots disc caps. $100-125

Six Shooter, 6 1/2"-long automatic with steel cylinder, made of nickel-plated cast iron in 1938 by Kilgore. Shoots disc caps. $60-80

Six Shooter, 6 1/2"-long automatic with steel cylinder, made of nickel-plated cast iron in 1938 by Kilgore. Shoots disc caps. $60-80

Spit Fire, 4 5/8" automatic made of nickel-plated cast iron in 1940 by Stevens. $40-60

Stevens 6 Shot, 6 1/4"-long automatic made of cast iron in 1932 by Stevens. $40-60

Stevens Repeater, 6 1/4" automatic made of nickel-plated cast iron in 1930 by Kilgore. $40-60

Sure Shot, 4 1/4"-long automatic made of painted cast iron in 1940 by Hubley. $30-40

25 Jr, 4 1/8"-long automatic made of cast iron in the late 1930s by Stevens. Note trigger variation. $30-40

25 Jr, 4 1/8"-long automatic made of nickel-plated cast iron in the late 1930s by Stevens. $30-40

25 Jr, 4 1/8"-long automatic made of cast iron in the late 1930s by Stevens. $30-40

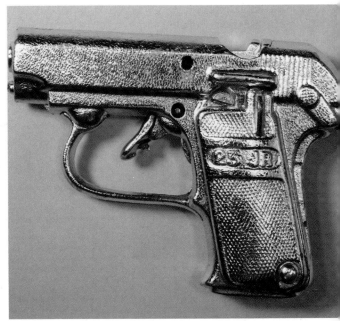

25 Jr, 4 1/8"-long automatic made of nickel-plated cast iron in the late 1930s by Stevens. Note the long bar variation. $30-40

25-50, 4 3/4" automatic made of nickel-plated cast iron in 1930 by Stevens. $60-80

Texan Jr, with horse 8 1/8"-long automatic made of nickel-plated cast iron in 1940 by Hubley. $100-125

Texan, with horse 9 1/4"-long automatic made of nickel-plated cast iron in 1940 by Hubley. $150-200

Texan, with star 9 1/4"-long automatic made of nickel-plated cast iron in 1940 by Hubley. $150-200

The Big Noise, 5 1/2"-long automatic made of nickel-plated cast iron in 1922 by Andes. $80-100

The Sheriff, 9" automatic made of nickel-plated cast iron in 1940 by Stevens. $125-150

Trooper Safety, 10 1/4"-long automatic made of nickel-plated cast iron in 1925 by Kilgore. $60-80

Western Boy, 7 3/4"-long automatic made of nickel-plated cast iron in 1940 by Stevens. $125-150

Winner, 4 3/4" automatic made of nickel-plated cast iron in 1940 by Hubley. $60-80

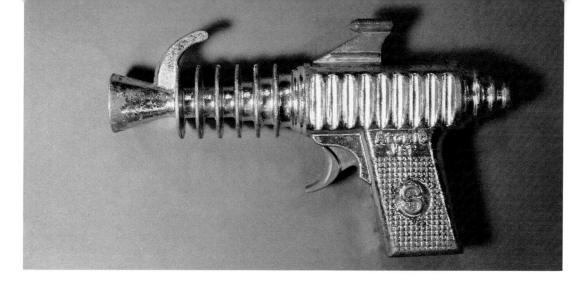

Atomic Jet Space Police Neutron, 7 3/4"-long automatic made of die-cast in 1949 by Blaster Cap Pistol. $150-200

Bobcat. 5"-long automatic made of nickel-plated die-cast by Kilgore. $20-30

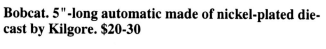

Brave, 6"-long single shot made of nickel-plated die-cast by Nichols. $20-30

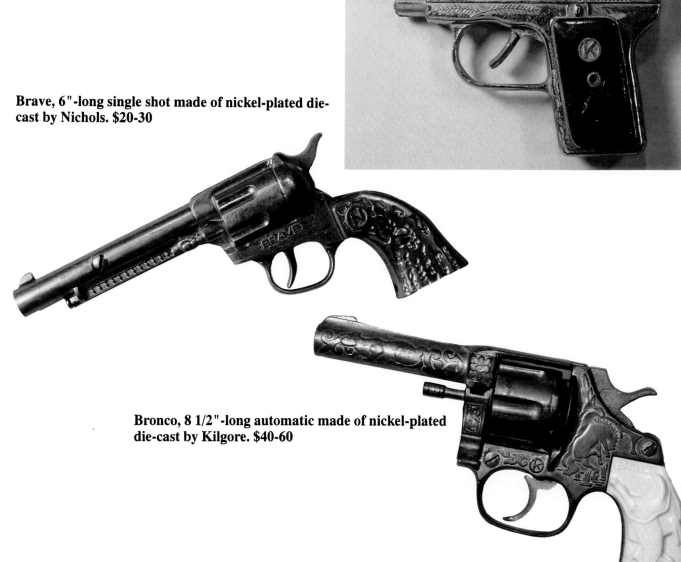

Bronco, 8 1/2"-long automatic made of nickel-plated die-cast by Kilgore. $40-60

Buck, 7 1/2"-long automatic made of nickel-plated die-cast by Kilgore. $40-60

Buckeroo, 8 1/4"-long automatic made of nickel-plated die-cast by Hubley. $30-40

Buffalo Bill, 10 1/4"-long automatic made of nickel-plated die-cast by Balantyone MFG. Chicago. $150-200

Buffalo Bill, 9"-long automatic made of nickel-plated die-cast by Hubley. $40-60

Bull Dog, 6 3/4"-long automatic made of nickel-plated die-cast. $20-30

Captain Hook, 11"-long single shot made of tin. $20-30

Champ, 5 3/4"-long automatic made of nickel-plated die-cast by Hubley. $20-30

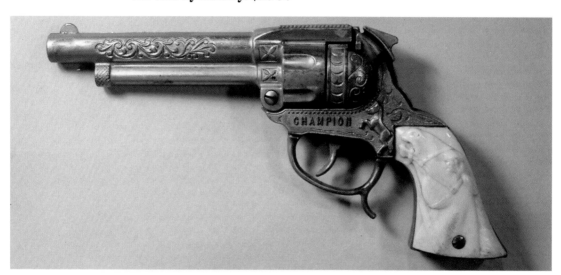

Champion, 9"-long automatic made of nickel-plated die-cast by Leslie Henry. $40-60

Cheyenne Shooter, 10"-long automatic made of die-cast by Hamilton. $20-30

Chief, 7 1/2"-long single shot made of nickel-plated die-cast by Hubley. $20-30

Clip 50, 4 1/2"-long automatic made of Bakelite and steel by Kilgore. $60-80

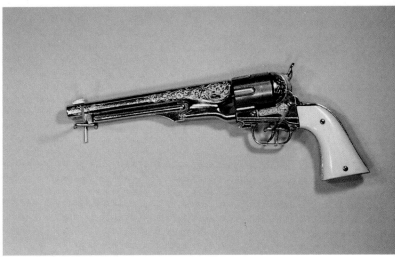
Colt 44, 14 1/2"-long automatic made of nickel-plated die-cast by Hubley. $80-100

Colt, 4"-long single shot made of nickel-plated die-cast. $30-40

Cowboy, 12 1/4"-long automatic made of die-cast by Hubley. $80-100

Cowboy Jr, 9 1/2"-long automatic made of nickel-plated die-cast by Hubley. $60-80

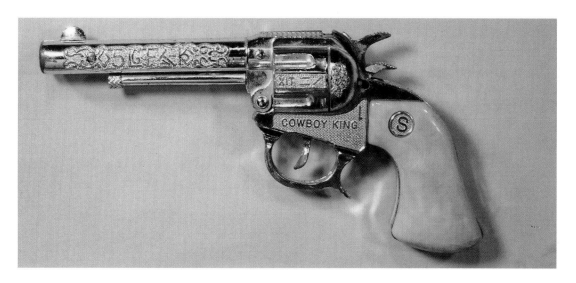

Cowboy King, 9"-long automatic made of nickel-plated die-cast by Stevens. $20-30

Cowpoke Jr, 8 1/4"-long automatic made of nickel-plated die-cast by Lone Star. $40-60

Cowpoke Jr, 8"-long automatic made of nickel-plated die-cast by Lone Star. $30-40

Daisy, 6 1/4"-long single shot made of painted die-cast by Daisy. $20-30

Dagger Derringer, 7 3/4"-long automatic made of nickel plated die-cast by Hubley. $80-100

Coyote, 8 3/4"-long automatic made of nickel-plated die-cast by Hubley. $40-60

Daisy, 8 3/4"-long automatic made of nickel-plated die-cast by Daisy. $40-60

Daisy, 10 1/4"-long automatic made of painted die-cast by Daisy MFG. Co. $40-60

Daisy, 6 1/4"-long single shot made of painted die-cast by Daisy. $20-30

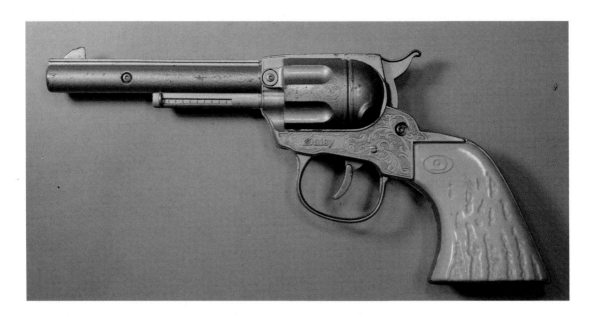

Daisy, 10"-long single shot made of nickel-plated die-cast by Daisy. $30-40

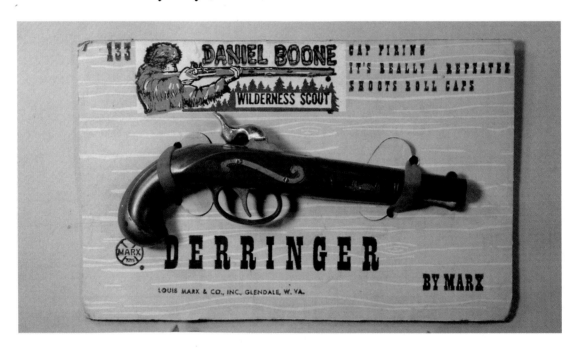

Daniel Boone, 8 1/4"-long single shot made of plastic and die-cast by Marx. $60-80

Davy Crockett, 12"-long automatic made of die-cast by Schmidt. $200-300

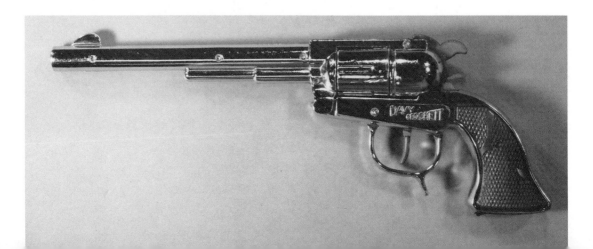

Deputy, 11"-long automatic made of die-cast by Hubley. $30-40

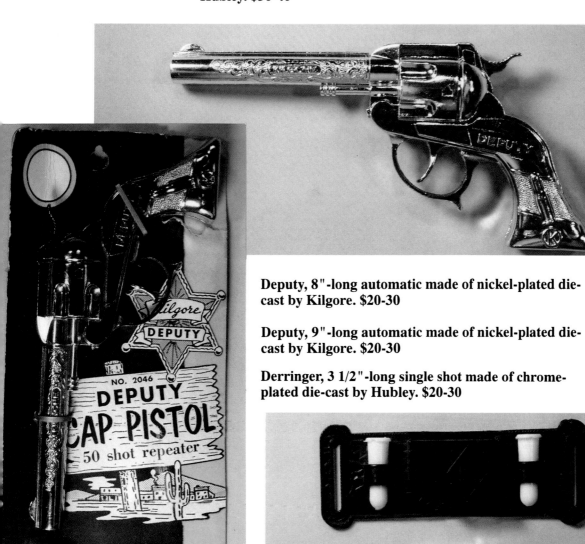

Deputy, 8"-long automatic made of nickel-plated die-cast by Kilgore. $20-30

Deputy, 9"-long automatic made of nickel-plated die-cast by Kilgore. $20-30

Derringer, 3 1/2"-long single shot made of chrome-plated die-cast by Hubley. $20-30

Detective 250, 6"-long automatic made of die-cast by Nichols. $20-30

Detective, 5 1/2"-long automatic made of nickel-plated die-cast by Leslie Henry. $20-30

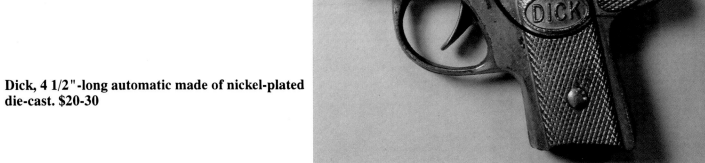

Dick, 4 1/2"-long automatic made of nickel-plated die-cast. $20-30

Dick, 5 1/4"-long automatic made of die-cast. $20-30

Dragnet, 6 1/2"-long automatic made of die-cast by Knickerbocker. $60-80

Dragnet, 7 1/2"-long automatic made of painted die-cast by Knickerbocker. $60-80

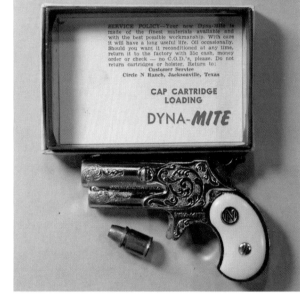

Dyna - Mite, 3 1/2"-long single shot with original box made of nickel-plated die-cast by Nichols. $20-30

Eagle, 8 1/2"-long automatic with rivet in grip made of nickel-plated die-cast by Kilgore. $80-100

Fanner 50, 11 1/2"-long automatic made of painted die-cast by Mattel. $50-70

Eagle, 8 1/2"-long automatic made of nickel-plated die-cast by Kilgore. $80-100

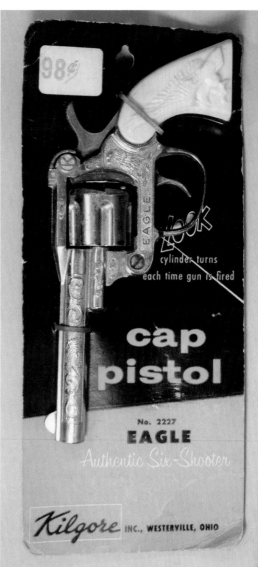

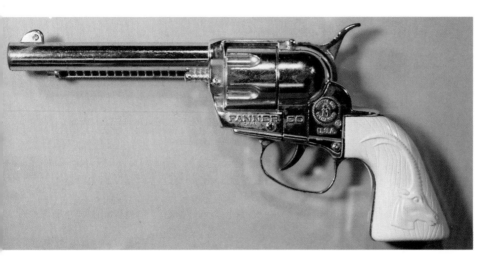

Fanner 50, 11 1/2"-long automatic made of nickel-plated die-cast by Mattel. $60-80

45 Smoker, 11 1/2"-long single shot made of painted die-cast by Product Eng. Co. $40-60

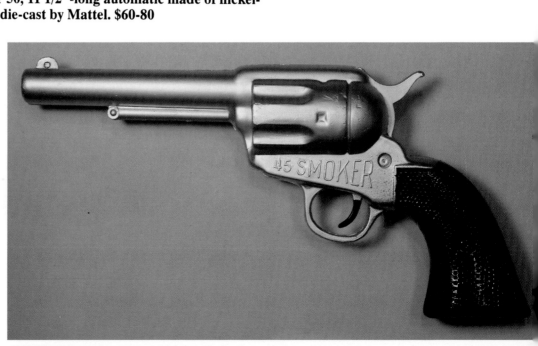

Frontier Smoker, 10 1/2"-long automatic made of nickel-plated die-cast. $40-60

G-Boy, 7 1/2"-long automatic made of painted die-cast by Acme Novelty. $20-30

Gabriel, 9"-long automatic made of nickel-plated die-cast by Hubley. $20-30

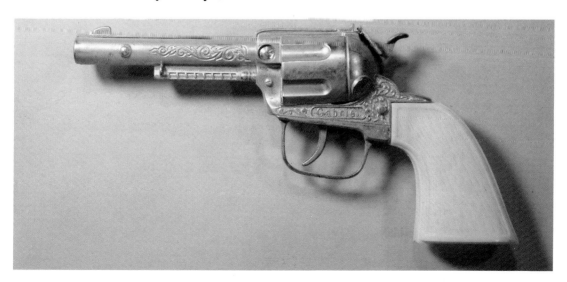

Gabriel, 9"-long automatic made of nickel-plated die-cast by Hubley. $30-40

Gene Aurty, 8 1/4"-long automatic made of nickel-plated die-cast by Kilgore. $125-150

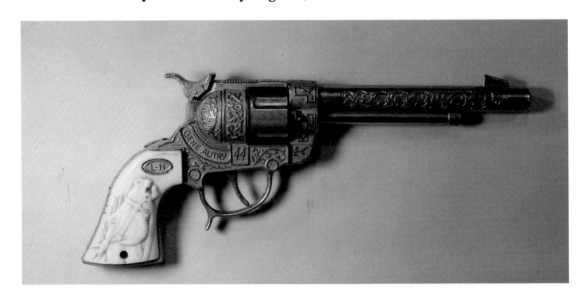

Gene Autry 44, 11 1/4"-long automatic made of die-cast by Leslie Henry. $150-175

Gene Autry, 8 1/8"-long automatic made of die-cast. $100-125

Gene Autry, 9 3/4"-long automatic made of nickel-plated die-cast by Kilgore. $125-150

Golden anniversary, The Rebel 8 3/4"-long automatic made of gold plated die-cast by Leslie Henry. $30-40

Grizzly, 10 1/2"-long automatic made of nickel-plated die-cast by Kilgore. $80-100

Hawkeye, 4 3/4"-long automatic made of nickel-plated die-cast by Kilgore. $20-30

Hawk, 5 3/4"-long automatic made of nickel-plated die-cast by Hubley. $40-60

Hide Away, 4"-long automatic made of die-cast. $10-20

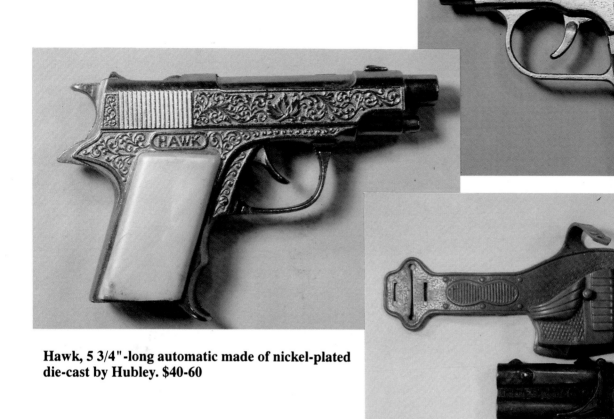

Hopalong Cassidy, 9 1/2"-long automatic made of nickel plated die-cast by Wyandotte. $80-100

Hopalong Cassidy, 9 3/4"-long automatic made of nickel plated die-cast by Geo. Schmidt. $125-150

Hopalong Cassidy, 9"-long automatic made of die-cast by Wyandotte. $80-100

Hub, 6 1/4"-long single shot made of die-cast in 1940 by Hubley. $20-30

Hubley, 10 1/2"-long automatic made of nickel-plated die-cast by Hubley. $20-30

Hubley, 7 1/4"-long automatic made of gold plated die-cast by Hubley. $100-125

Hubley, 7 3/4"-long automatic made of nickel-plated die-cast by Hubley. $20-30

Hubley, 9"-long automatic made of nickel-plated die-cast by Hubley. $20-30

Hubley, horse head grip 7 1/2"-long automatic made of nickel-plated die-cast by Hubley. $60-80

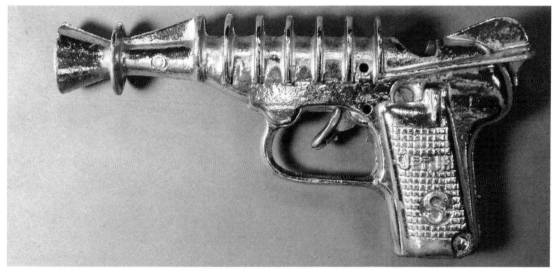

Jet Jr, 6 3/8" automatic made of die-cast and cast Iron in 1945 by Stevens. $150-200

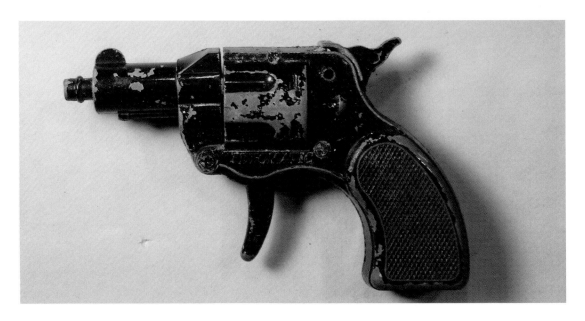

Jetomatic, 5 1/4"-long single shot cap gun and water gun made of die-cast. $30-40

Keen Shot Jr., 5"-long single shot made of painted die-cast by Callen MFG. Co. $20-30

Kentucky Pistol, model 1729, 13 1/2"-long single shot made of die-cast and wood by Parris Mfg. $30-40

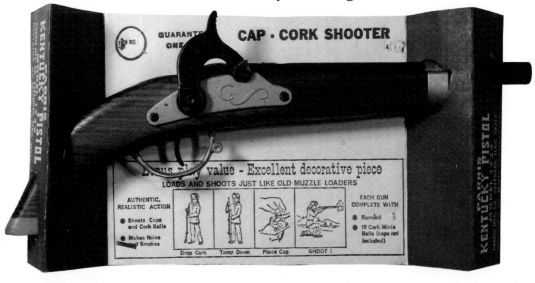

Kit Carson, 10"-long automatic made of nickel-plated die-cast by Kilgore. $60-80

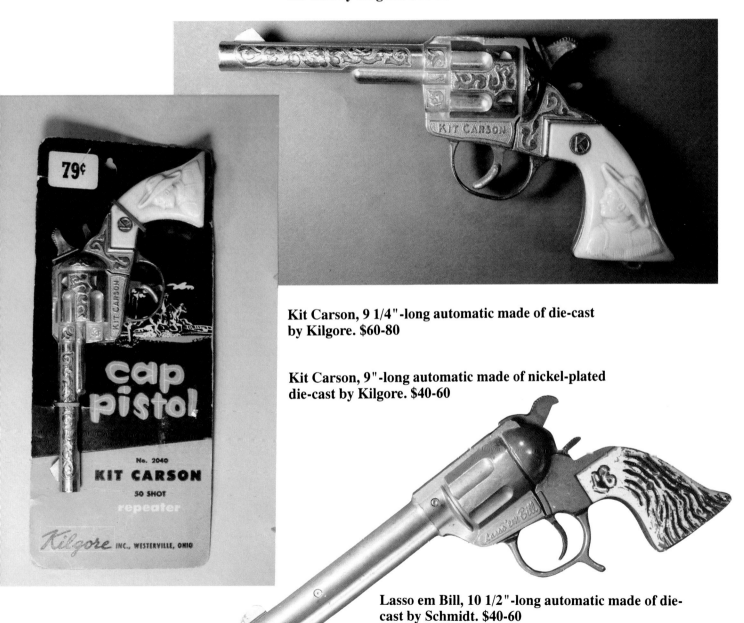

Kit Carson, 9 1/4"-long automatic made of die-cast by Kilgore. $60-80

Kit Carson, 9"-long automatic made of nickel-plated die-cast by Kilgore. $40-60

Lasso em Bill, 10 1/2"-long automatic made of die-cast by Schmidt. $40-60

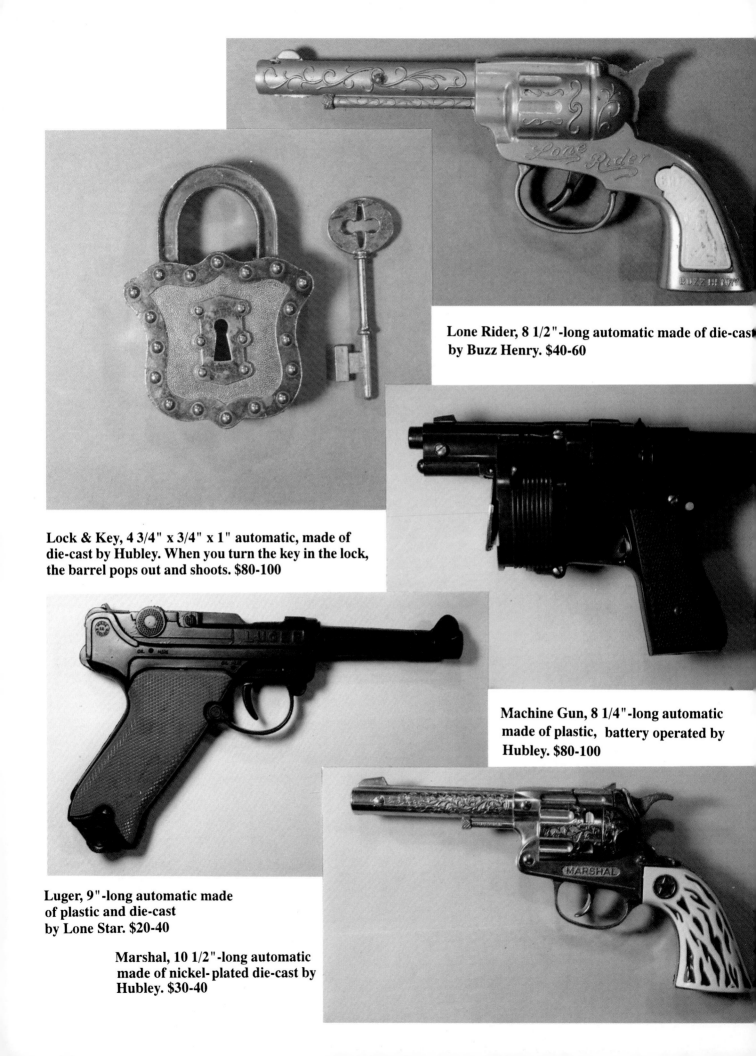

Lone Rider, 8 1/2"-long automatic made of die-cast by Buzz Henry. $40-60

Lock & Key, 4 3/4" x 3/4" x 1" automatic, made of die-cast by Hubley. When you turn the key in the lock, the barrel pops out and shoots. $80-100

Machine Gun, 8 1/4"-long automatic made of plastic, battery operated by Hubley. $80-100

Luger, 9"-long automatic made of plastic and die-cast by Lone Star. $20-40

Marshal, 10 1/2"-long automatic made of nickel-plated die-cast by Hubley. $30-40

Marshal, 10 1/2"-long automatic made of nickel-plated die-cast by Hubley. $40-60

Marshal, 9"-long automatic made of nickel-plated die-cast made by Hubley. $30-40

Maverick 45, 11 3/4"-long automatic made of bronze plated die-cast by Leslie Henry. $60-80

Military, 7 1/2"-long automatic made of nickel-plated die-cast by Halio. $20-40

Mustang 500, 12 3/4"-long automatic made of nickel-plated die-cast by Nichols. $60-80

Mustang, 9 1/2"-long automatic made of chrome-plated die-cast by Kilgore. $60-80

National Metal, 11"-long automatic made of chrome-plated die-cast by National Metal Plastic Co. $80-100

No Name, 11 1/2"-long automatic made of nickel-plated die-cast by Hubley. $20-30

One Shot, 4 1/2"-long single shot made of nickel-plated die-cast by Kilgore. $20-30

Pal, 6"-long automatic made of nickel-plated die-cast by Hubley. $20-30

Paladin, 9 3/4"-long automatic made of nickel-plated die-cast by Leslie Henry. $60-80

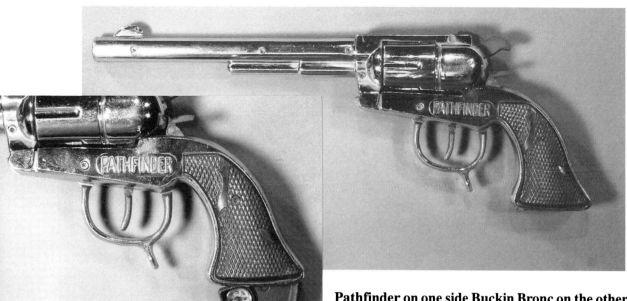

Pathfinder on one side Buckin Bronc on the other side. Has a compass in the handle, 12"-long automatic made of die-cast in the late 1940s and 1950s by Schmidt. $150-200

Pet, 6 1/4"-long automatic made of nickel-plated die-cast by Hubley. $20-30

Pioneer, 10 1/4"-long automatic made of die-cast by Hubley. $30-40

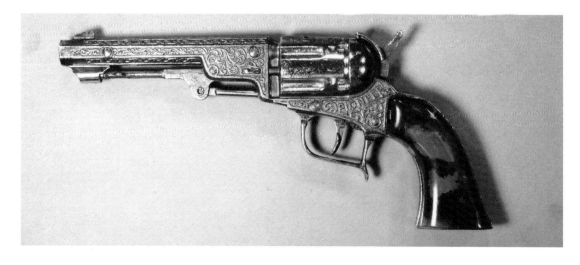

Pirate. 9 3/8"-long automatic made of nickel-plated die-cast in 1940 by Hubley. $80-100

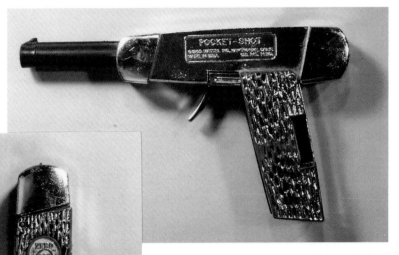

Pocket-Shot, 4 1/2" X 1 1/2" automatic made of diecast and plastic by Mattel. $30-40

Police, 8 3/4"-long automatic made of painted steel by Meldon. $20-30

Pony Boy, 10 1/2"-long automatic made of nickel-plated die-cast by Actoy. $20-30

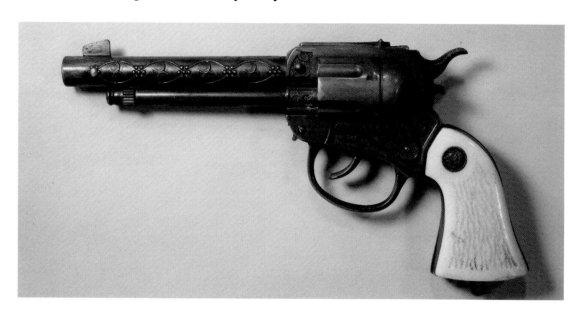

Pony Boy, 10 1/4"-long automatic made of bronze plated die-cast by Actoy. $40-60

Pony Boy, 10"-long automatic made of nickel-plated die-cast by Actoy. $30-40

Pony, 7 1/4"-long single shot made of nickel-plated die-cast by Nichols. $20-30

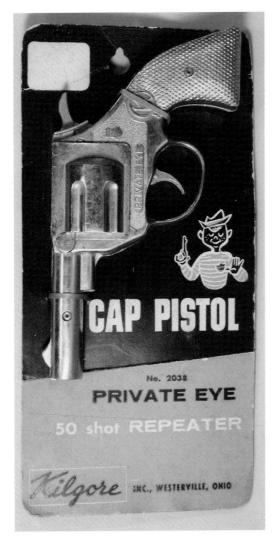

Private Eye, 7"-long automatic made of nickel-plated die-cast by Kilgore. $30-40

Pup, 5 3/4"-long single shot made of die-cast by Hubley. $20-30

Rancher, 8"-long automatic made of nickel-plated die-cast by Wyandotte. $30-40

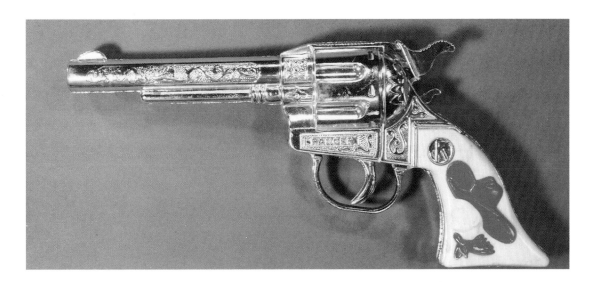

**Ranger, 9 3/4" automatic made of die-cast by Kilgore.
$30-40**

**Red Ranger, 8"-long automatic made of nickel-plated
die-cast by Wyandotte. $30-40**

**Red Ranger, 9 1/2"-long automatic made of die-cast
by Wyandotte. $30-40**

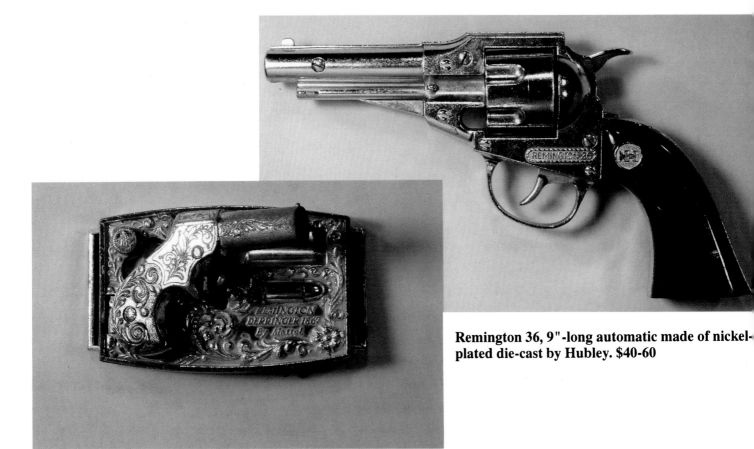

Remington 36, 9"-long automatic made of nickel-plated die-cast by Hubley. $40-60

Remington Derringer 1867 4"-long automatic made of chrome-plated die-cast by Mattell. $20-30

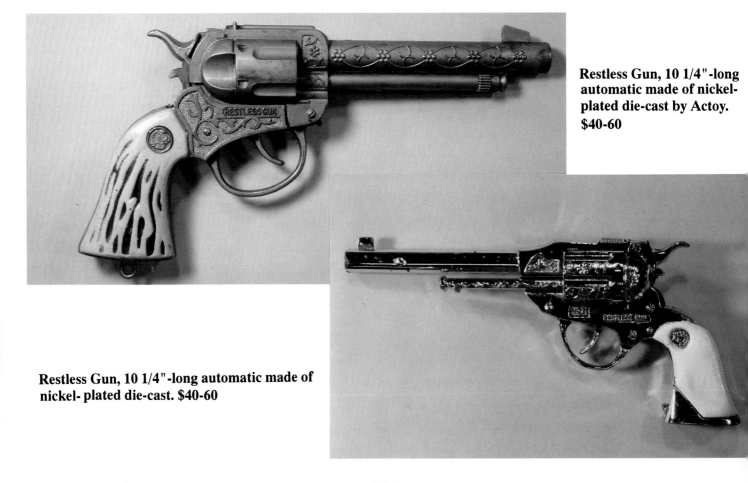

Restless Gun, 10 1/4"-long automatic made of nickel-plated die-cast by Actoy. $40-60

Restless Gun, 10 1/4"-long automatic made of nickel- plated die-cast. $40-60

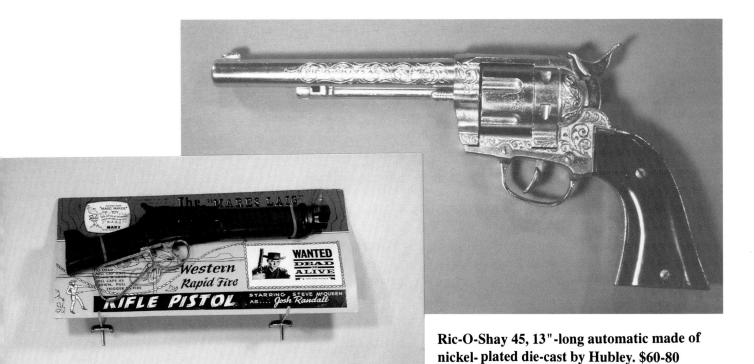

Ric-O-Shay 45, 13"-long automatic made of nickel- plated die-cast by Hubley. $60-80

Rifle Pistol, 13"-long made of plastic and die-cast by Marx. $100-120

Rodeo, 8 3/4"-long automatic made of nickel-plated die-cast by Hubley. $30-40

Ring Gun, shoots caps single shot by Harry C. Bjorklund made in the 1950s. $40-60

Rodeo, 7 1/2"-long single shot made of nickel-plated die-cast in 1940 by Hubley. $20-30

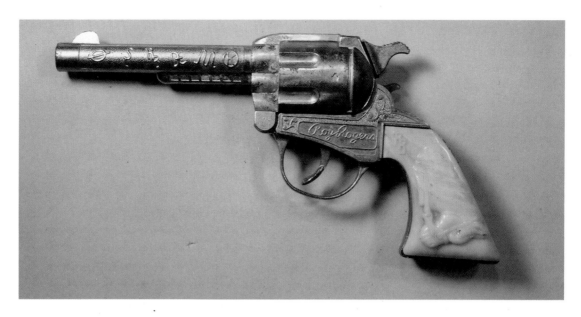

Roy Rogers, 8 1/2"-long automatic made of chrome-plated die-cast by Kilgore. $80-100

Roy Rogers, 10 1/2"-long automatic made of chrome-plated die-cast by Classy. $100-125

Roy Rogers, 10 1/2"-long automatic made of die-cast by Nichols. $125-150

Roy Rogers, 10 1/2"-long automatic made of chrome-plated die-cast by Schmidt. $125-150

Roy Rogers, 9 3/4"-long automatic made of chrome-plated die-cast by Geo. Schmidt. $125-150

Roy Rogers, 9"-long automatic made of nickel-plated die-cast made by Schmidt. $125-150

Ruf Rider, 10 3/4"-long automatic made of nickel-plated die-cast made by Latco. $40-60

Secret Agent junior, 5"-long repeater made of nickel-plated die-cast by Hamilton. $20-30

Secret Agent, 5 3/4"-long automatic made of nickel-plated die-cast by Hamilton. $20-30

Shootin Shell 45, 12"-long automatic made of nickel-plated die-cast by Mattel. $100-150

Shootin Shell Snub Nose 38, 6 1/2"-long automatic made of nickel-plated die-cast by Nichols. $40-60

Silver Pony, 8 1/4"-long single shot made of nickel-plated die-cast by Nichols. $20-30

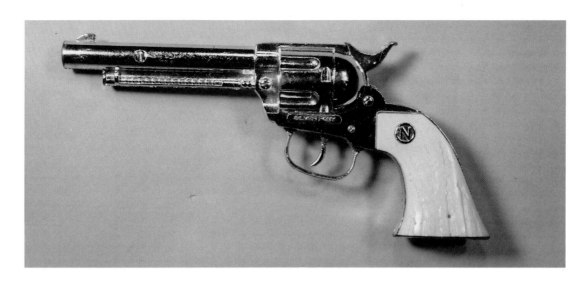

Silver Pony, 8 3/4"-long automatic made of nickel-plated die-cast by Nichols. $20-30

Smoky Joe, with Texas-longhorn on grips, 9 3/4"-long automatic made of die-cast by Schmidt. $60-80

Smoky, 6"-long automatic made of nickel-plated die-cast by Hubley. $20-30

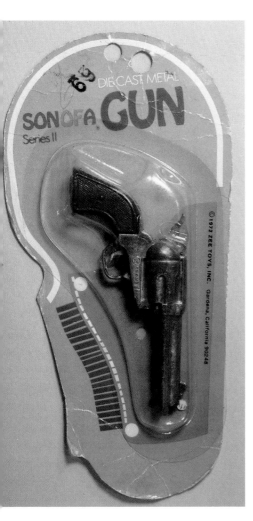

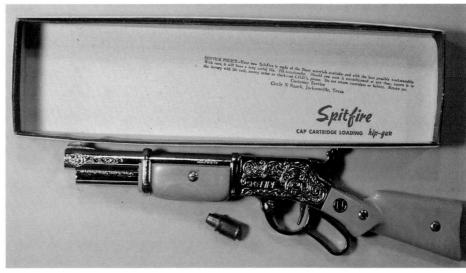

Son of A Gun, 4 1/2"-long automatic made of die-cast in 1972 by Zee Toys. $20-30

Spitfire, 9"-long single shot with original box made of nickel-plated die-cast by Nichols. $30-40

Stallion 32, 8 3/4"-long automatic made of nickel-plated die-cast by Nichols. $40-60

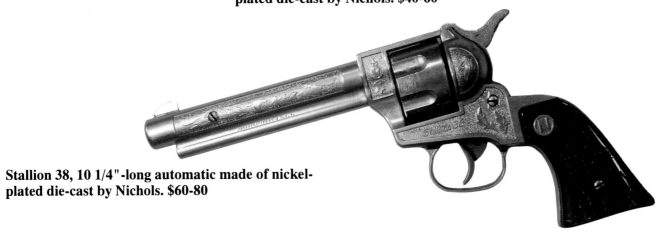

Stallion 38, 10 1/4"-long automatic made of nickel-plated die-cast by Nichols. $60-80

Stallion 41-40, 11"-long automatic made of nickel-plated die-cast by Nichols. $150-175

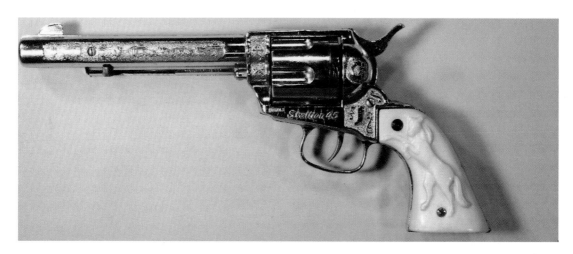

Stallion 45, 12 1/2"-long automatic horse on grip, made of chrome-plated die-cast by Nichols. $150-175

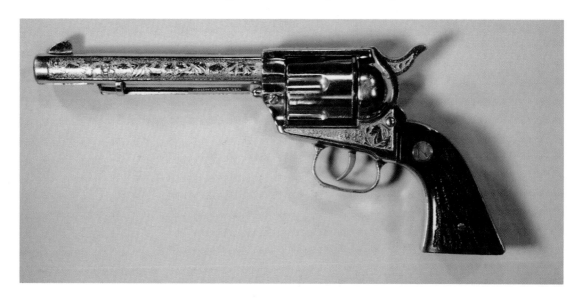

Stallion 45, 12 1/2"-long automatic made of chrome-plated die-cast by Nichols. $150-175

Stallion, 10"-long automatic made of nickel-plated die-cast by Nichols. $60-80

Star, 7"-long automatic made of nickel-plated die-cast by Kilgore. $20-30

Strato Gun, 9 1/2"-long automatic made of nickel-plated die-cast in Detroit Mi. $150-200

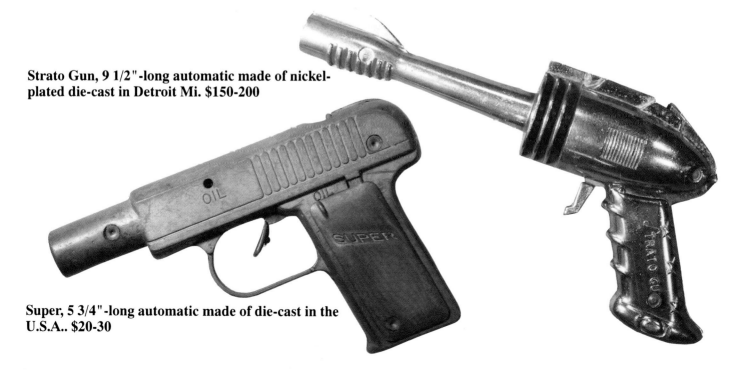

Super, 5 3/4"-long automatic made of die-cast in the U.S.A.. $20-30

Sure Shot, 8"-long automatic made of nickel-plated die-cast by Hubley. $40-60

Sure Shot, 9"-long single shot made of nickel-plated die-cast by Hubley. $40-60

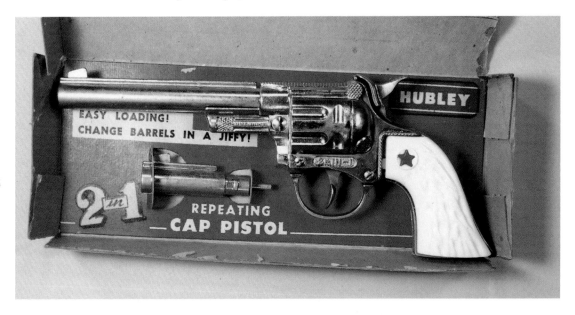

2 in 1, 9"-long automatic made of nickel-plated die-cast by Hubley. $60-80

250 Shot, 7"-long automatic made of die-cast by Actoy. $40-60

357 Magnum, 7"-long automatic made of painted die-cast by Daisy. $20-30

Tex, 8 1/2"-long automatic made of nickel-plated die-cast by Hubley. $40-60

Texan 38, 11 1/4"-long automatic made of die-cast by Hubley. $80-100

Texan Jr. 9 1/2"-long automatic made of nickel-plated die-cast made by Hubley. $40-60

Texan Jr, 8 1/8"-long automatic made of gold plated die-cast in 1940 by Hubley. $60-80

Texan Jr. 10"-long automatic made of nickel-plated die-cast made by Hubley. $40-60

Texan Jr, 10"-long automatic made of chrome-plated die-cast in 1950 by Hubley. $40-60

Texan Jr, 10 1/4"-long automatic made of chrome-plated die-cast by Hubley. $40-60

Texas, 9 3/4"-long automatic made of nickel-plated die-cast. $40-60

Texas, 6 3/4"-long automatic made of nickel-plated die-cast by Leslie Henry. $20-30

Texas Kid, 7"-long automatic made of nickel-plated die-cast by Straco. $30-40

Texas Kid, 10 1/4"-long automatic made of nickel-plated die-cast by Hubley. $30-40

Texas Ranger, 10"-long automatic made of nickel-plated die-cast by Leslie Henry. $40-60

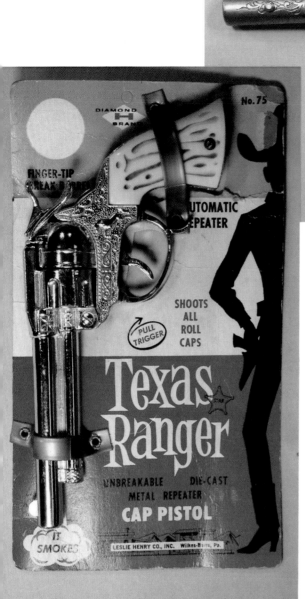

The Unexcelled Automatic, 7 1/2", made of tin. $20-30

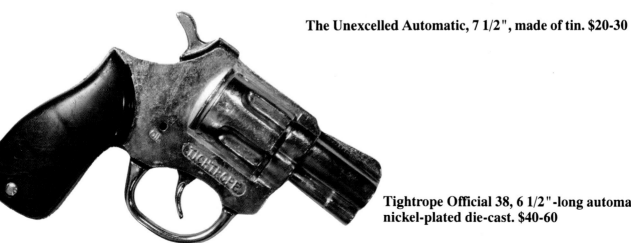

Tightrope Official 38, 6 1/2"-long automatic made of nickel-plated die-cast. $40-60

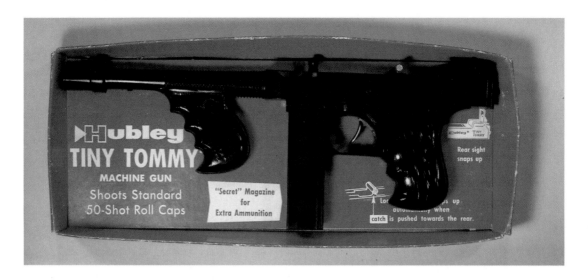

Tiny Tommy, 10 1/2"-long automatic made of plastic and die-cast by Hubley. $20-30

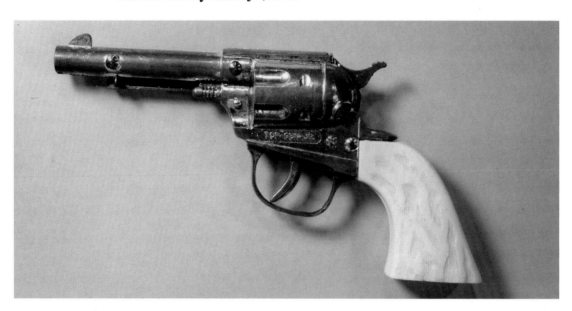

Top-Gun Jr., 8"-long automatic made of die-cast. $20-30

Trigger, 9"-long automatic made of nickel-plated die-cast by Stevens. $40-60

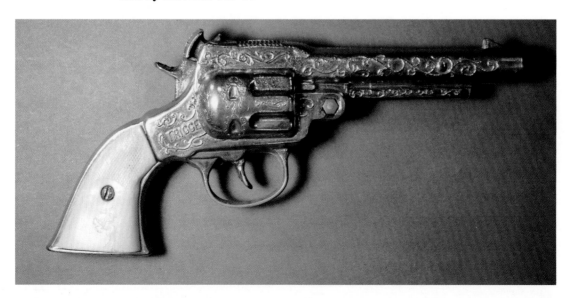

Trooper, 7"-long automatic made of nickel-plated die-cast by Hubley. $20-30

Trooper, 6 1/2"-long automatic made of nickel-plated die-cast by Hubley. $20-30

Trooper, 6 /2"-long automatic made of die-cast by Gabriel. $20-30

ULI, 7"-long single shot made of tin in Germany. $20-30

U.S. 45, 7 1/4"-long automatic made of plastic in 1975 by Chemtoy. $20-30

Unxed, 5"-long automatic made of nickel-plated tin. $20-30

Victory Gatler, 4 1/2"-long automatic with four spoke trigger or crank made of die-cast in 1925 by Victory Spark. $150-200

Western Man 9 1/2"-long automatic made of nickel-plated die-cast. $20-30

Western Man 9"-long automatic made of nickel-plated die-cast. $20-30

Western, 9 1/2"-long automatic made of chrome-plated die-cast by Hubley. $60-80

Western, 9 1/2"-long automatic made of chrome-plated die-cast by Hubley. $30-40

Wyatt Earp Buntline Special, 11 1/2"-long automatic made of nickel-plated die-cast by Hubley. $80-100

Y&, 9 1/4"-long automatic made of nickel-plated die-cast by Wyandotte. $40-60

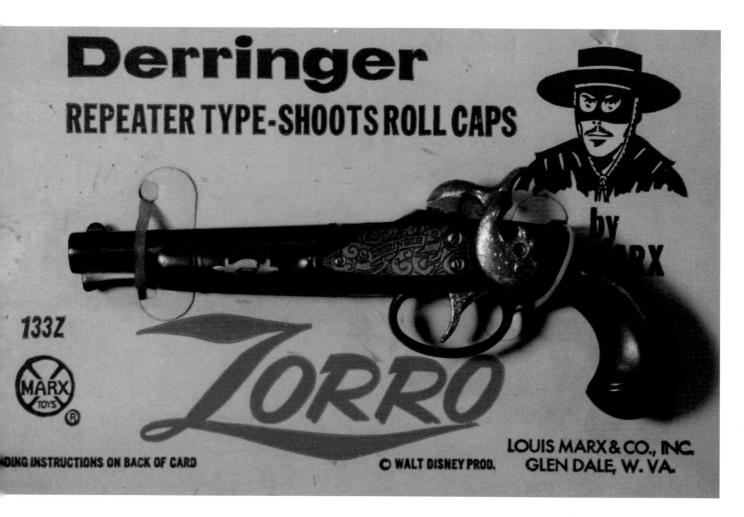

Zorro Derringer, 8 1/2"-long single shot made of plastic and die-cast by Marx. $60-80

RIFLES

Machine-gun, 13 1/2"-long automatic made of plastic and die-cast by Mattel. $60-80

Machine-gun, 23"-long automatic made of plastic and die-cast by Mattel. $60-80

Machine-gun, 16 1/2"-long automatic made of plastic and die-cast made in 1957 by Mattel. $60-80

Machine-gun, 20 3/4"-long automatic made of plastic and die-cast by Knickerbocker. $40-60

Big Game Rifle, 36"-long automatic made of plastic and die-cast by Marx. $60-80

Colt Six Shooter, 31"-long automatic made of plastic and die-cast by Mattel. $60-80

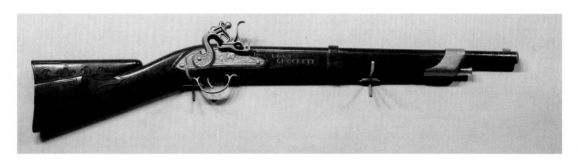

Davy Crockett, 24 1/2"-long automatic made of plastic and die-cast by Hubley. $150-175

Davy Crockett, 34 1/4"-long automatic made of plastic and die-cast by Walt Disney. $150-175

Frontier Fighter, 26"-long automatic made of plastic and die-cast by Larami. $60-80

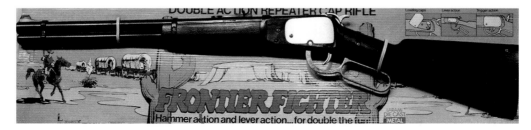

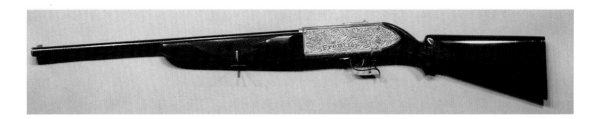

Frontier, 34 1/2"-long automatic made of plastic and die-cast with metal receiver. $60-80

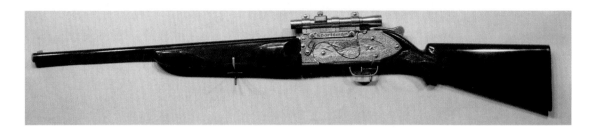

Sportsman, 34 1/2"-long automatic made of plastic and die-cast by Hubley. $60-80

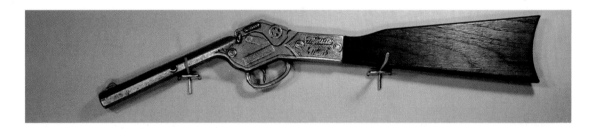

Minute Man, 20" automatic made of nickel plated cast Iron in 1936 by Kilgore. $200-300

Roy Rogers, 26"-long automatic made of plastic and die-cast by Marx. $125-150

Scout Rifle, 32 1/2"-long automatic made of plastic and die-cast by Hubley. $100-125

Frontier, 34 1/2"-long automatic made of plastic and die-cast with plastic receiver. $60-80

The Lone Ranger, 34"-long automatic made of plastic and die-cast by Marx. $125-150

The Rifleman, 32 1/2"-long automatic made of plastic and die-cast by Hubley. $150-200

Winchester Saddle Gun, 33"-long automatic made of plastic and die-cast by Mattel. $60-80

Zero, Agent Zero radio rifle, 9 3/8"-long automatic made of plastic in 1964 by Mattel. $80-100

Zero, Agent Zero radio rifle, 9 3/8"-long automatic made of plastic in 1964 by Mattel.

HOLSTERS

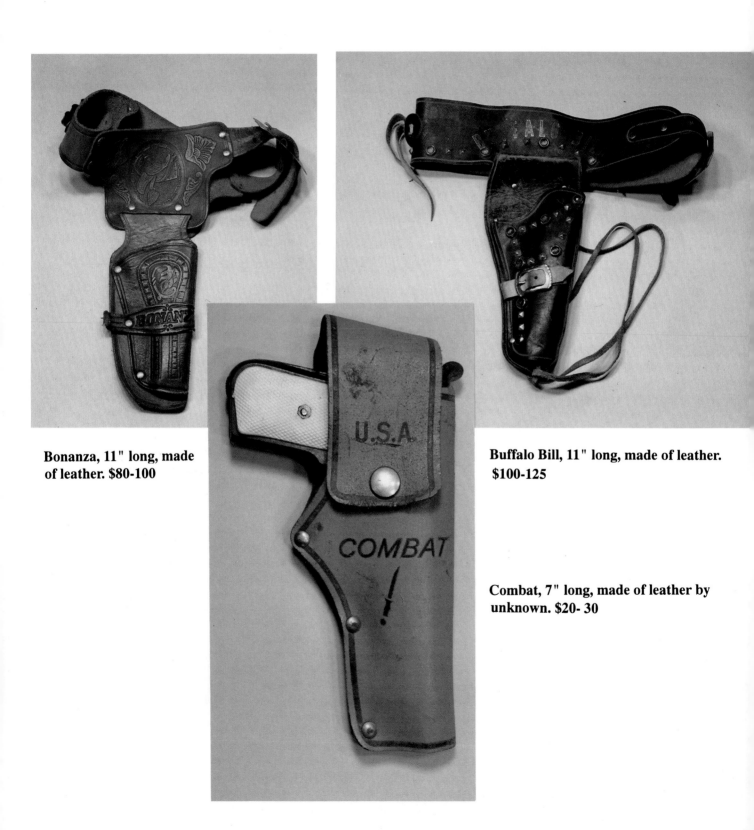

Bonanza, 11" long, made of leather. $80-100

Buffalo Bill, 11" long, made of leather. $100-125

Combat, 7" long, made of leather by unknown. $20-30

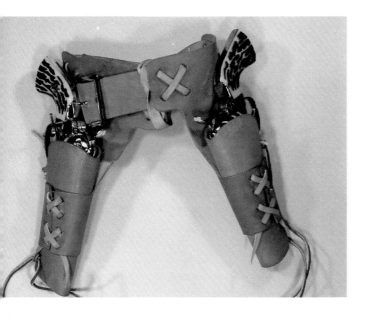

Fanner 50 set, 11 1/4" long, made of leather by Mattel. $100-150

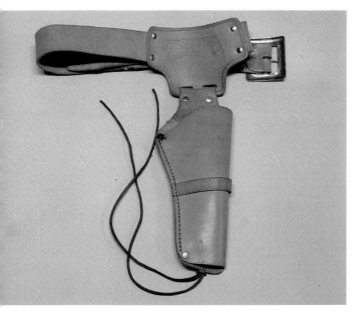

Fanner 50, 11" long, made of leather by Mattel. $60-80

Gene Autry, 10" long, made of leather. $40-60

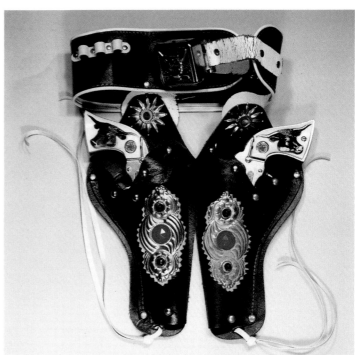

Gunsmoke set, 11" long, made of leather. $125-150

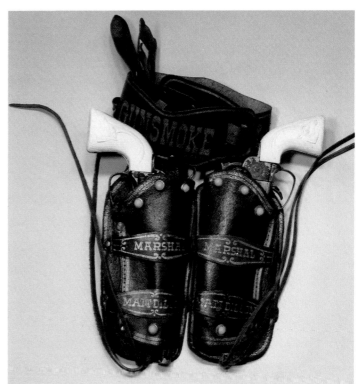

Gunsmoke, 10" long, made of leather by unknown. $150-175

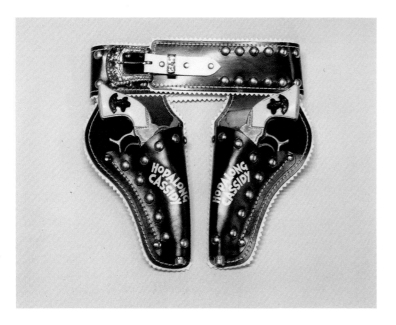

Hopalong Cassidy reproduction, 9" long, made of leather. $100-125

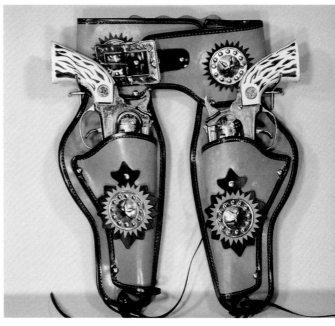

Indian Leather set, 11" long, made of leather by Indian Leather Co. $125-150

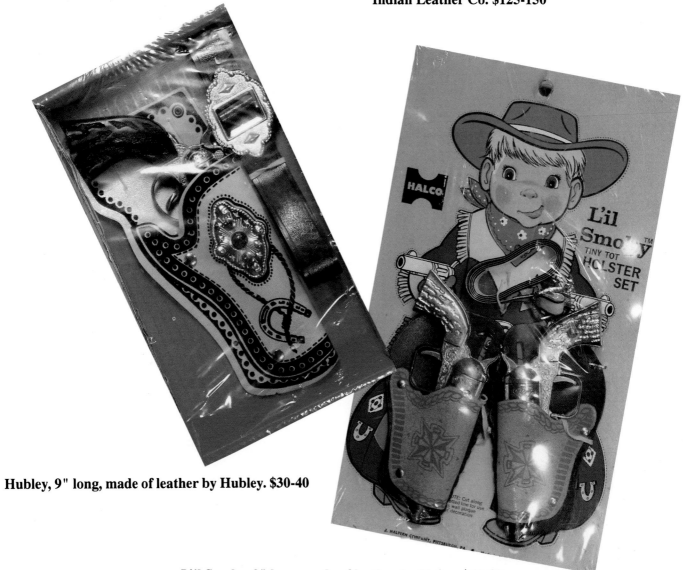

Hubley, 9" long, made of leather by Hubley. $30-40

L'il Smoky, 3" long, made of leather by Halco. $60-80

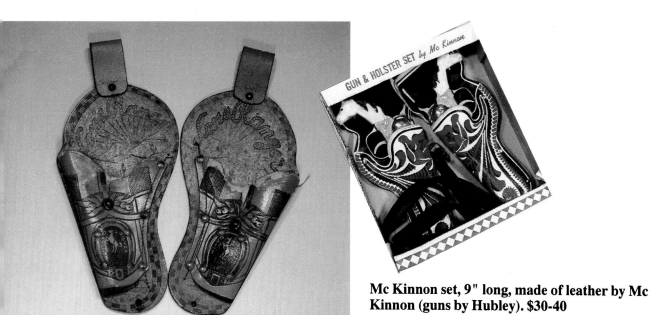

Mc Kinnon set, 9" long, made of leather by Mc Kinnon (guns by Hubley). $30-40

Lone Ranger set, 12" long, made of leather. $100-125

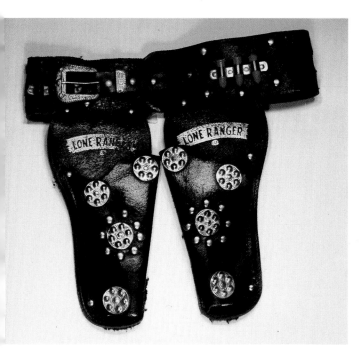

Lone Ranger, 8" long, made of leather. $100-150

No Name, 10" long, made of leather by unknown. $100-150

Lone Ranger set, 10" long, made of leather. $80-100

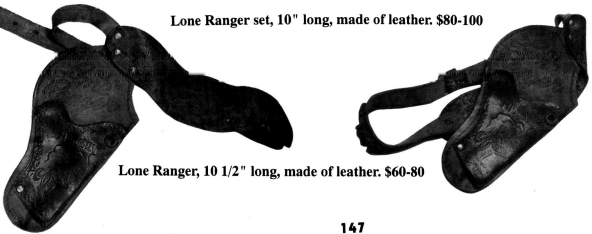

Lone Ranger, 10 1/2" long, made of leather. $60-80

No Name, 10" long, made of leather by unknown. $80-100

No Name, 10" long, made of leather by unknown. $80-100

No Name, 11" long, made of leather by unknown. $150-200

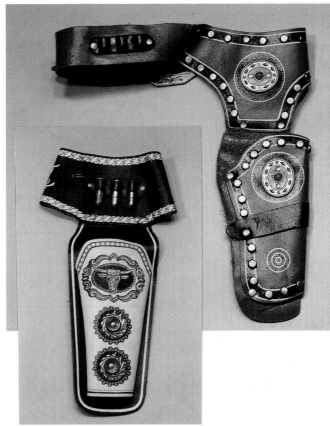

No Name, 10" long, made of leather by unknown. $40-60

No Name, 10" long, made of leather by unknown. $20-30

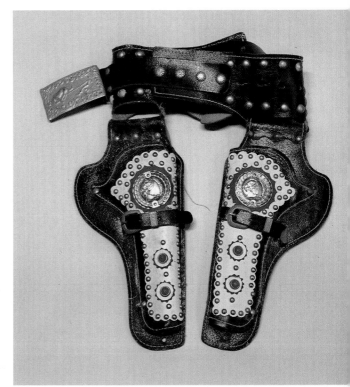

No Name, 11" long, made of leather by unknown. $60-80

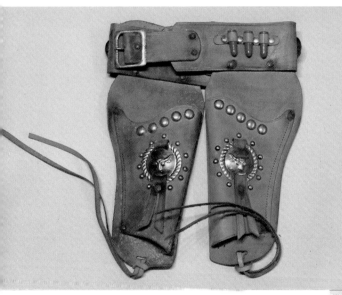

No Name, 12" long, made of leather by unknown. $40-60

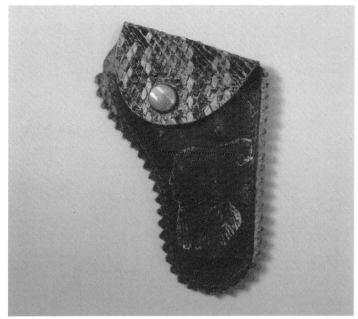

No Name, 3" long, made of leather by Hubley. $10-20

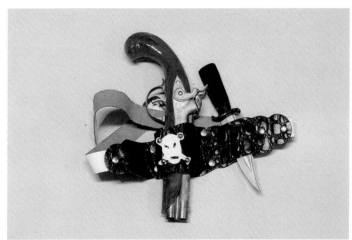

No Name, 3" long, made of leather by unknown. $40-60

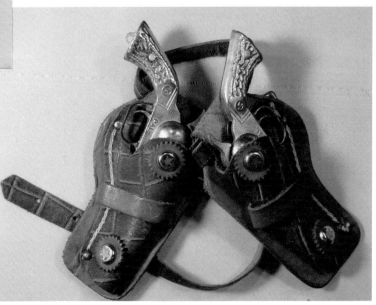

No Name, 4" long, made of leather by unknown. $40-60

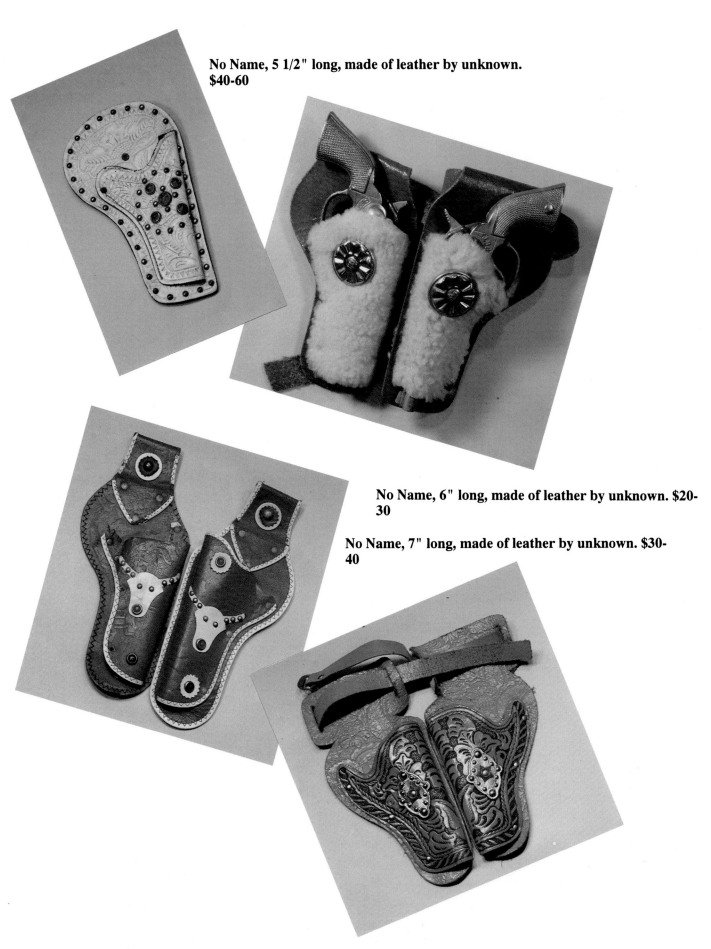

No Name, 5 1/2" long, made of leather by unknown. $40-60

No Name, 6" long, made of leather by unknown. $20-30

No Name, 7" long, made of leather by unknown. $30-40

No Name, 7" long, made of leather by unknown. $40-60

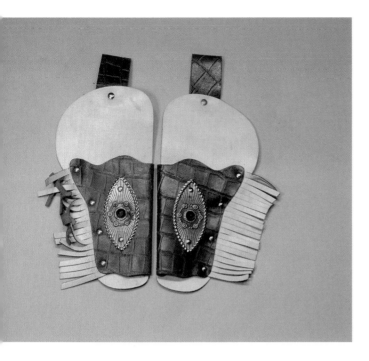

No Name, 7" long, made of leather by unknown. $20-30

No Name, 8" long, made of leather by unknown. $20-30

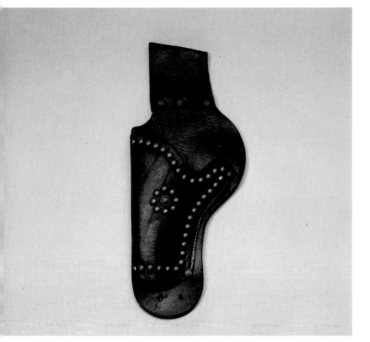

No Name, 7" long, made of leather by unknown. $20-30

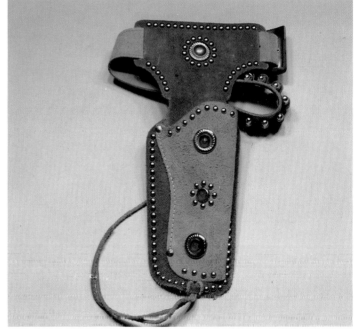

No Name, 8" long, made of leather by unknown. $40-60

No Name, 8" long, made of leather by unknown. $80-100

No Name, 8" long, made of leather by unknown. $60-80

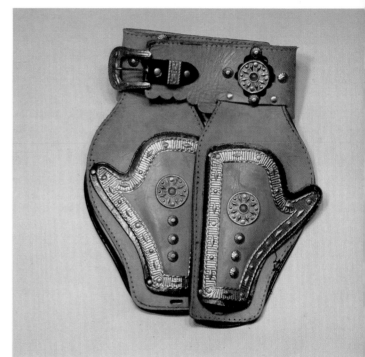

No Name, 8" long, made of leather by unknown. $30-40

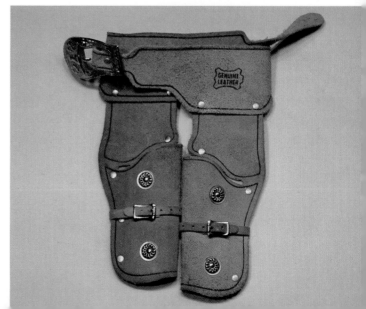

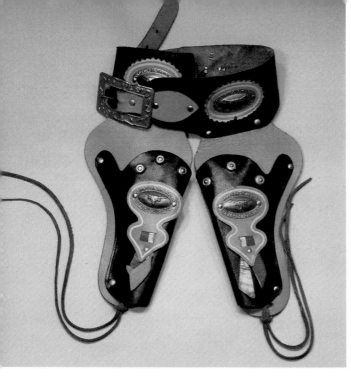

No Name, 8" long, made of leather by unknown. $80-100

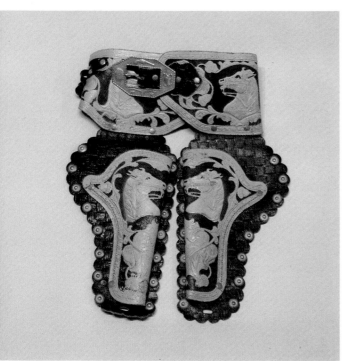

No Name, 8" long, made of leather by unknown. $100-125

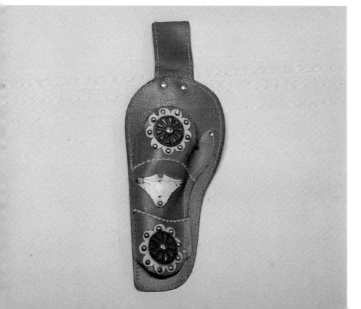

No Name, 9" long, made of leather by unknown. $20-30

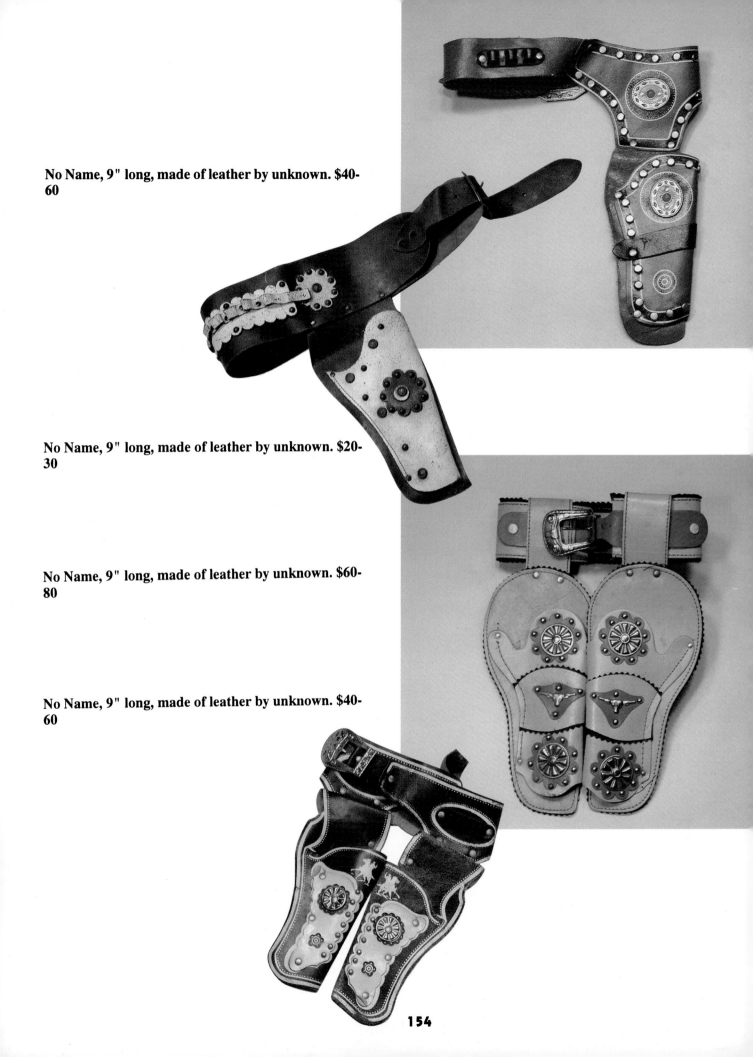

No Name, 9" long, made of leather by unknown. $40-60

No Name, 9" long, made of leather by unknown. $20-30

No Name, 9" long, made of leather by unknown. $60-80

No Name, 9" long, made of leather by unknown. $40-60

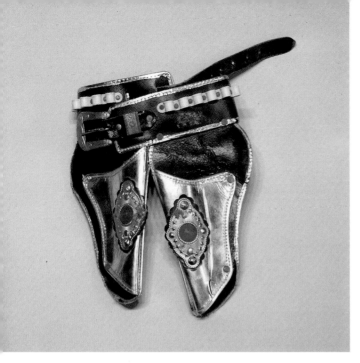

No Name, 9" long, made of leather by unknown. $60-80

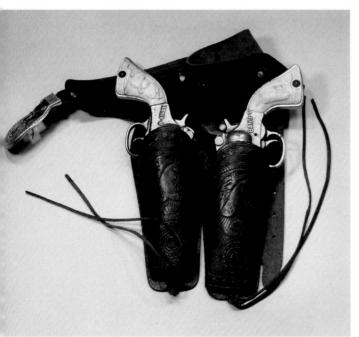

No Name, 9" long, made of leather by unknown. $80-100

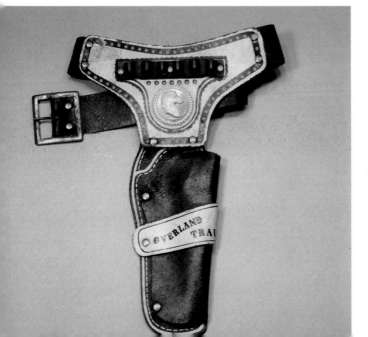

Overland Train, 12" long, made of leather. $20-30

Paladin set, 12 1/2" long, made of leather. $150-200

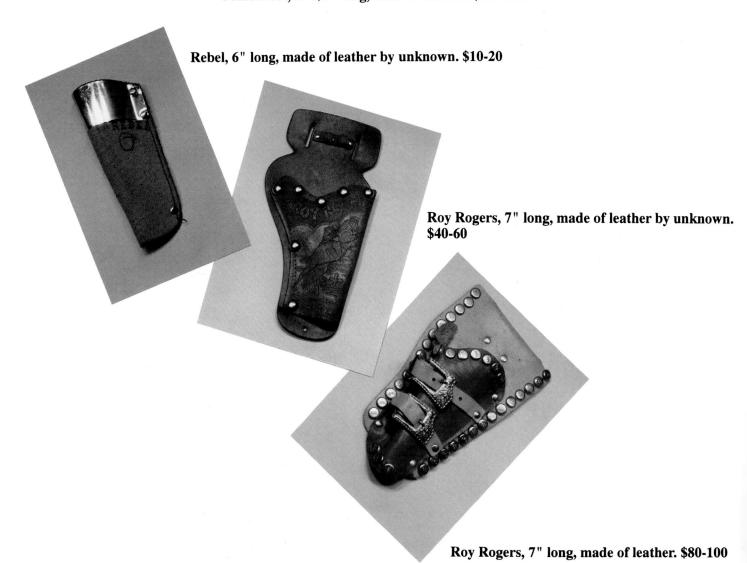

Rebel, 6" long, made of leather by unknown. $10-20

Roy Rogers, 7" long, made of leather by unknown. $40-60

Roy Rogers, 7" long, made of leather. $80-100

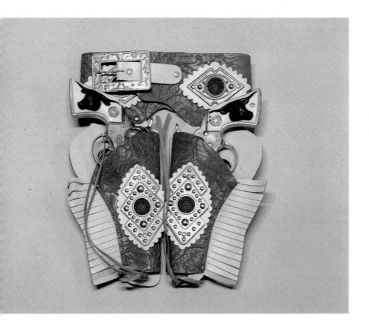

Texan, 8" long, made of leather by Halco. $80-100

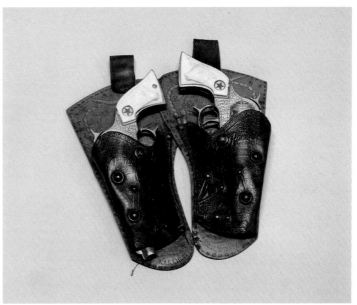

Texas Ranger, 10" long, made of leather by M.A.H. in 1946. $60-80

Tuck A Way, 3" long, made of leather. $20-30

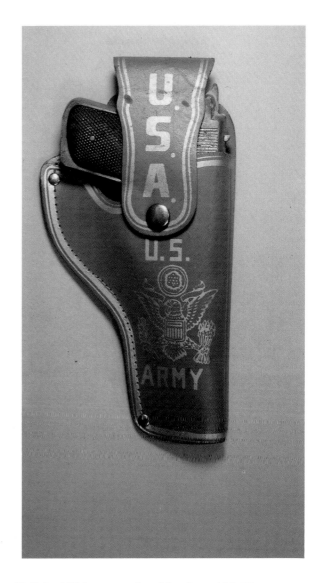

U.S.A. 10" long, made of leather. $20-30

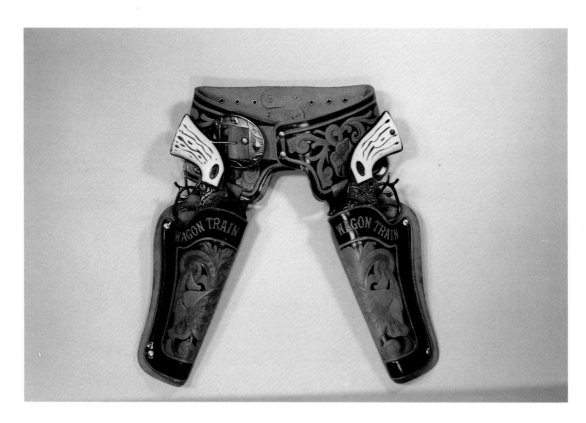

Wagon Train, 11 1/4" long, made of leather by Leslie Henry. $150-200

Western Frontier set, 11" long, made of leather by Jaco Novelty Co. $60-80

No Name set, 11" long, made of leather. $150-175

USING PATENT NUMBERS

Patent Numbers can be a great help in telling when the object was first made.

DATE	Patent #	DATE	Patent #	DATE	Patent #
1836	1	1889	395305	1942	2268540
1837	110	1890	418655	1943	2307007
1838	546	1891	443987	1944	2338061
1839	1061	1892	466315	1945	2366154
1840	1465	1893	483976	1946	2391856
1841	1923	1894	511744	1947	2413675
1842	2413	1895	531619	1948	2433824
1843	2901	1896	552502	1949	2457797
1844	3395	1897	574369	1950	2492944
1845	3873	1898	596467	1951	2536016
1846	4348	1899	616871	1952	2580379
1847	4914	1900	640167	1953	2624046
1848	5409	1901	664827	1954	2664562
1849	5993	1902	690385	1955	2698434
1850	6961	1903	717521	1956	2728913
1851	7865	1904	758567	1957	2775762
1852	8622	1905	778834	1958	2818567
1853	9512	1906	808618	1959	2868973
1854	10358	1907	839799	1960	2919443
1855	12117	1908	875679	1961	2966681
1856	14009	1909	908430	1962	3015103
1857	16324	1910	945010	1963	3070801
1858	19010	1911	980178	1964	3116487
1859	22477	1912	1013095	1965	3163365
1860	26642	1913	1049326	1966	3226729
1861	31005	1914	1083267	1967	3295143
1862	34045	1915	1125212	1968	3360800
1863	37266	1916	1166419	1969	3419907
1864	41047	1917	1210389	1970	3487470
1865	45685	1918	1251458	1971	3551909
1866	51784	1919	1290027	1972	3633214
1867	60658	1920	1326899	1973	3707729
1868	72959	1921	1354054	1974	3781914
1869	85503	1922	1401948	1975	3858241
1870	98460	1923	1440352	1976	3930271
1871	110617	1924	1478996	1977	4000520
1872	122304	1925	1521590	1978	4065812
1873	134504	1926	1568040	1979	4131952
1874	146120	1927	1612700	1980	4180876
1875	158350	1928	1654521	1981	4242757
1876	171641	1929	1696897	1982	4308622
1877	185813	1930	1742181	1983	4366579
1878	198753	1931	1787424	1984	4423523
1879	211078	1932	1839190	1985	4490885
1880	223211	1933	1892663	1986	4562596
1881	236137	1934	1941449	1987	4633526
1882	251685	1935	1985678	1988	4716594
1883	269320	1936	2026516	1989	4794652
1884	291016	1937	2066309	1990	4890335
1885	310163	1938	2104004	1991	4980927
1886	333494	1939	2142080	1992	5077836
1887	355291	1940	2185170		
1888	375720	1941	2227418		